UNIT

WJEC A2 G4

Geography

Sustainability

David Burtenshaw and
Sue Warn

PHILIP ALLAN
UPDATES

First published in 2010 by Philip Allan Updates, an imprint of Hodder Education, an Hachette UK company, Market Place, Deddington, Oxfordshire OX15 0SE

Orders

Bookpoint Ltd, 130 Milton Park, Abingdon, Oxfordshire OX14 4SB
tel: 01235 827827
fax: 01235 400401
e-mail: education@bookpoint.co.uk

Lines are open 9.00 a.m.–5.00 p.m., Monday to Saturday, with a 24-hour message answering service. You can also order through the Philip Allan Updates website: **www.philipallan.co.uk**

© David Burtenshaw, Sue Warn 2010

ISBN 978-1-4441-1085-2

First printed 2010

Impression number 5 4 3 2

Year 2014 2013 2012 2011

This guide has been written specifically to support students preparing for the WJEC A2 Geography Unit G4 examination. The content has been neither approved nor endorsed by WJEC and remains the sole responsibility of the author.

Printed by MPG Books, Bodmin

Hachette UK's policy is to use papers that are natural, renewable and recyclable products and made from wood grown in sustainable forests. The logging and manufacturing processes are expected to conform to the environmental regulations of the country of origin.

P01779

Contents

Questions and Answers

Introduction

About this guide

The purpose of this guide is to help you understand what is required to achieve a good grade in **Unit G4: Sustainability**. The full contents of the specification are available on the WJEC website: **www.wjec.co.uk**.

To help with your studies the guide is divided into three major sections. This **Introduction** explains the structure of the guide. It also provides guidance on how to approach the Unit G4 examination and its pre-release resources.

The **Content Guidance** section gives a guide to the key questions for each of the four themes. It also provides some exemplar material that can be used when answering questions. The exam paper is divided into two sections. Section A contains four questions (Questions 1–3, worth 10 marks each and Question 4, worth 25 marks). All questions relate to the themes of the pre-release resources and your own studies of the themes. The pre-release resources are in a Resource Folder that you will receive prior to the examination. This cannot be taken into the examination. You will receive a fresh copy of the Resource Folder with the question paper. It is anticipated that the resources will cover *at least two of the four themes* in the specification. Your responses to Section A can be completed from the resources and from your own studies. However, to rely solely on your own studies rather than the Resource Folder is a dangerous course to take. Some questions will be wider than the resources but even then your response ought to make use of the resources. Section B contains one question (25 marks) that relates to one or more of the themes. It expects you to draw on knowledge from your studies of Unit G4 and of the other units in the specification.

Unit G4 is a synoptic paper that draws on your studies from the G4 specification as well as knowledge built up throughout your geographical studies at A-level. One of the skills that you should have acquired is the ability to identify connections between the different aspects of geography. In addition, you should be able to apply your knowledge and understanding from other parts of your studies to the unfamiliar context that the Resource Folder presents. This paper is one that contributes to the award of an A* grade (Unit G3 is the other). Therefore, a good understanding of the content of the specification, together with an ability to relate topics to one another in the heat of the examination, is essential if you want to achieve the highest award.

Unit G4 is an assessment that relies on you writing essays in continuous prose. To help develop your essay-writing skills, the Questions and Answers section includes examples of the types of question that you will see in the examination. Sample answers are provided together with examiner comments on how to tackle each question and how to improve the answers.

The Unit G4 exam

There is no choice of questions — all questions are compulsory. The questions do not have to be answered in order. If you are more confident with Section B, you may answer Question 5 first. However, it is best to answer Questions 1–4 in order because they develop a theme.

Timing

The exam lasts 1 hour 45 minutes, in which time you must answer all five questions. The question paper provides guidance on the amount of time that you should devote to each question (see the Questions and Answers section). The table below gives you an approximate idea of the timings that you should try to follow.

Tasks in examination room	Time in minutes
Section A	
Read all five questions	2
Draw up a resource/question grid (see page XX)	4
Plan your response to Question 1	1
Answer Question 1 (10 marks)	8
Plan your response to Question 2	1
Answer Question 2 (10 marks)	8
Plan your response to Question 3.	1
Answer Question 3 (10 marks)	8
Plan your response to Question 4	2
Answer Question 4 (25 marks)	30
Review your answers to Questions 1–4	2
Section A total time	**1 hour 7 minutes**
Section B	
Plan your response to Question 5	5
Answer Question 5 (25 marks)	31
Review your answer to Question 5	2
Section B total time	**38 minutes**
Total time	**1 hour 45 minutes**

Quality of written communication

No marks are allocated specifically for the quality of your writing. However, you should try to use correct punctuation and grammar, especially in Questions 4 and 5 where you are expected to respond in essay style. Your responses should be structured and logical, include an introduction and a conclusion, and use appropriate geographical terminology. If you are dyslexic or have other special needs that inhibit your ability to write, you should make sure that your school has sought special consideration on your behalf.

Geographical terms and examples

It is essential that you know the key terms associated with sustainability. Since this paper is synoptic, it relies on knowledge from Units G1, G2 and G3, so you must revisit your own list of key terms. It is also useful to build up your own memory bank of simple, effective maps and diagrams. Finally, you will need your own examples to illustrate the points that you make in your answers. You will come across examples and diagrams in this book; remember that you can impress by the use of different, relevant examples that you have found rather than the standard ones in your textbooks.

Managing questions in a minute

All questions have three major components:
- **command words** such as outline, discuss, evaluate, assess
- **subject matter**
- **locations** — whether specified or left for you to choose. Geography answers are best if they include details of places. The Resource Folder may focus on a set of places, regions or countries and you should aim to make good use of these

Use highlighter pens in different colours to emphasise the command words, the subject matter and the location. In addition, your responses to the questions in Section A will gain more credit if you demonstrate that you know and understand what the resources are showing. Therefore, build up a reference grid for the resources when you first look at the examination paper, as suggested on pages 95–96.

Command words
- **Assess** — weigh up the importance of the topic. There will be a number of possible explanations and you need to give the main ones and then say which you tend to favour.
- **Compare** asks you to point out the similarities, although many versions of such questions expect some contrast. **Contrast** strictly expects the differences. **Compare and contrast** requires both similarities and differences.
- **Classify** expects you to group ideas or phenomena or explanatory variables into categories.
- **Describe** expects you to show what you know is occurring or what a map or diagram shows. It expects you to know what a process does, what happens, where it occurs and when it occurs. Who or what causes a process and whom it may affect can be part of a description.
- **Discuss** and **Discuss the assertion** both expect you to build up an argument about an issue and to present more than one side of the evidence. This type

of question requires you to reach a conclusion. Discussion will include both description and explanation and a summary at the end.

- **Evaluate** — you should try to state a viewpoint, having looked at the overall explanations for an issue for which you have considered the evidence. Credit will be given for the justification of the view that you take. Expect there to be more than one explanation.
- **Examine** asks you to investigate in detail, giving evidence both for and against a point of view or an opinion.
- **Explain** expects you to say why and how something occurs. It might involve some description but do not rely on description as a means of explanation.
- **With the aid of a diagram** means that you must draw one. It must be labelled and the axes on a graph identified.
- **Identify** requires you to provide listed points that are supported by examples.
- **Justify** is asking you to state why one opinion or explanation is better than another.
- **Outline** is asking you to state the main points or factors and expects more than two points with supporting examples.
- **To what extent** and **How far do you agree** both expect explanations for and against, together with a justification of the view that you favour.

Content Guidance

Unit G4: Sustainability consists of four themes:

- sustainable food supply
- sustainable water supply
- sustainable energy
- sustainable cities

You have to study *all* of them. Section A, which is based primarily on the Resource Folder, is anticipated to cover at least two of the four themes, whereas Section B will address another theme. Do not anticipate that a theme in Section B will be omitted because it has come up recently. The examiner may repeat a theme at any time.

The Content Guidance begins with an overview of sustainability, which can be applied to all four themes. Each theme in this guide has a general introductory section based on the sustainability quadrant (see page 12). You are then guided through the four key questions that you should have investigated. This book does not provide you with the synopticity required for the Unit G4 examination. You must use knowledge from your studies in the other three units in the geography specification.

Introducing sustainability

The classic definition of sustainable development comes from the Brundtland Report of 1987: development that meets the needs of the present without compromising the ability of future generations to meet their own needs.

It contains two key concepts:
- that of needs such as water, food and shelter which are essential for the survival of the world's poor
- that of limitations imposed by the state of technology and social organisation on the environment's ability to meet both present and future needs

This **futurity principle** can be summarised as 'not cheating on your children' by controlling the use of finite resources such as fossil fuels, minimising waste and pollution and protecting natural habitats. It is a beguilingly simple concept, embraced by leaders right across the political spectrum from Margaret Thatcher to Tony Blair to Bill Clinton.

Although the United Nations Environmental Programme (UNEP) report *Our Common Future* suggested that equity for all people, environmental protection and economic growth were all simultaneously possible, the reality is far more complex when considering sustainable food, water and energy supplies and cities.

Sustainability can be likened to a three-legged stool, where the legs are the environment, society and economy. A failure to address any one of these will result in the stool toppling over and the objective of sustainable development being thwarted.

Figure 1 The stool of sustainability

Over the last 20 years sustainability has been redefined and the concept expanded to include a number of key strands. Figure 2 shows the sustainability quadrant, which has four major facets:

- Futurity based on the Brundtland principle.
- Environment — eco-friendly 'greenness' based on principles developed following the Rio Summit of 1992.
- Public participation based on small-scale localised strategies (the Schumacher principle — small is beautiful) where people are empowered to take responsibility for their own development.
- Equity and social justice with pro-poor strategies — developed during the 2002 Sustainable Development Summit in Johannesburg.

Futurity	Environment
Present generations should leave future generations the ability to maintain present standards of living whether through natural or cultural capital. Therefore, we are entitled to use up finite natural resources only if we provide future generations with the know-how (through improved science, technology and social organisation) to maintain living standards from what is left. This is the Brundtland principle.	We should seek to preserve the integrity of ecosystems, both at the local level and globally at the scale of the biosphere, in order not to disrupt natural processes that are essential to the safeguarding of human life and to maintaining biodiversity. This includes the eco-friendly management of water, land, wildlife, forests etc.
Public participation	**Equity and social justice**
The public should be aware of, and participate in, the process of change towards sustainable development in line with Rio Summit (1992) Principle 10. Environmental issues are best handled with the participation of all concerned citizens. Each individual should have the information and opportunity to participate in the decision-making process (bottom-up).	This principle implies fair shares for all including the most disadvantaged, locally and globally. 'If there is a finite amount that we may consume or use beyond which we cannot go...then we must share what we already have far more than is currently the case. Equality of access to the world's global resources therefore must be the guiding principle, improving the lives of the poor.' (Johannesburg, 2002)

Figure 2 The sustainability quadrant

Therefore, many agencies have developed different definitions of sustainability in order to develop sustainable strategies. For example, for the Countryside Commission sustainable development means meeting four objectives:

- Social progress that recognises the needs of everyone.
- Effective protection of the environment.
- Prudent use of natural resources.
- Maintenance of high and stable levels of economic growth and employment.

Action towards achieving sustainability therefore has to happen at a number of scales, as shown in Table 1. Strategies need to be coordinated across the scales — hence the notion of 'think global, act local' or vice versa.

Table 1 Action towards sustainable development

Level	Examples of action
Internationally/ globally	• Protocols, treaties and conventions (e.g. Montreal IPCC). • International legislation — ITTO, Convention on International Trade in Endangered Species (CITES), Rio (Biodiversity/Agenda 21). • International organisations' aid (e.g. UN, World Bank). • International trade organisations (e.g. WTO, GATT, IMF).
Nationally	• Development of national policies and plans using frameworks. • Development of legislation and taxation. • National organisations — some governmental (e.g. Environmental Agency).
Non-governmental organisations (NGOs)	• All have plans for sustainable schemes (e.g. Oxfam, ActionAid).
Locally via Agenda 21	• Local authorities — educational and environmental policies. • International Council for Local Environmental Initiatives (ICLEI).

As you will find when investigating the four themes, there are a number of challenges that make sustainable development difficult to achieve. Table 2 summarises the enormity of the challenges.

Table 2 Challenges of sustainable development

Resources are essential to economic and social development — global energy and global water crises have resulted.
All production and consumption activities produce waste — cleaning up land, water and air pollution remains a challenge (some pollution builds up, such as arsenic).
Many of the challenges, such as enhanced global warming and biodiversity loss, are global but the world is split into an increasing patchwork of nation states with different interests.
Poverty remains an enormous brake on the concept of sustainable development as the root cause of many problems can be linked to poverty.
World inequality is a major challenge with the wealthy North and poor South.
Previous unsustainable development has led to huge problems — the environment cannot cope if there is an enormous human cost.
Questions of responsibility and response are major challenges — although the polluter should pay, assigning responsibility is hard.
The power to respond is sometimes not politically, socially or economically possible.
Questions of sovereignty of individual countries can prevent global solutions as nation states pursue their own interests.

Theme 1:
Sustainable food supply

1.1 The issues

In 2006, although around 850 million people were living with **food insecurity** — defined as 'an unreliability of food supply', i.e. hunger — overall, the world had sufficient supplies of food to feed all its peoples. Technological advances in agriculture such as the Green Revolution meant that globally more food was grown at lower cost and globalisation, which led to improved communication and transport systems, facilitated the movement of food more cheaply over long distances. Globally there was clearly no **food availability deficit** (a facet of sustainability) although localised famines did occur, largely as a result of human factors.

The main reason for hunger and malnutrition to persist is poverty, which leads to a **food entitlement deficit** whereby poor countries, communities and individuals cannot afford to buy the available food. These people are more often landless, frequently living in households with little wage-earning capacity, living as AIDS orphans or unemployed as recent urban migrants. Some 96% of these hungry people were found in sub-Saharan Africa or south Asia. The other 4% lived in non-oil rich former Soviet Union countries such as Tadjikistan or in pockets within advanced industrialised countries. The USA, for example, shows marked inequalities, with the poorest people concentrated in the rural Appalachians or inner-urban areas of cities such as Detroit.

However, in 2008 the world food supply situation worsened considerably leading to a global food crisis. The Director of the UN World Food Programme wrote 'over 4 million more people join the ranks of the world's hungry each year, in spite of targets such as Millennium Development Goals (MDGs) which aim to halve the number of people experiencing food insecurity by 2015. Currently we are facing the tightest food supplies in recent history. For the world's most vulnerable, food is simply being priced out of their reach. This could lead to 1 billion people being hungry, i.e. a worsening situation of dwindling supplies and soaring food prices.'

The reasons for this are complex. There are three key interrelated issues:
- Threats to food availability, especially from short-term climate change and also from competition from biofuels for arable land.
- Rising demands for existing food supplies generated by rising consumption as a result of the nutritional transition of increasingly affluent populations in India and China and other Far Eastern countries, which leads to an increased demand for meat, dairy products and processed foods and subsequently cereals.
- The world's population is rising, with an estimated 2.5 billion more mouths to feed by 2050.

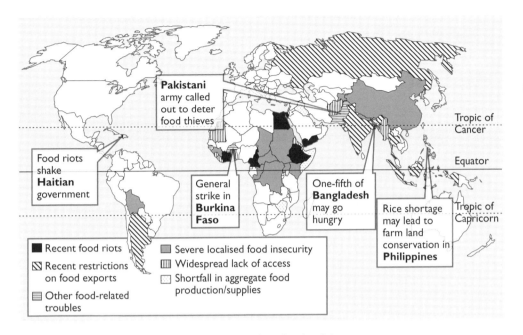

Figure 3 Countries facing food crisis 2007–08

At the same time, the **obesity pandemic** associated with developed nations highlights a world of unequal access to sustainable food supplies (see page 20).

Defining sustainable food supplies: eat well and save the planet?

The issue of defining sustainable food supplies is highly complex because this sustainability should be embedded at all stages of the food supply chain, from production to transportation and storage to consumption. Therefore, the definition needs to embrace issues such as what is the best type of farming system: eco-centric or techno-centric; controversies of food miles; and ethical consumerism with many players such as agrochemical and supermarket TNCs and governments having an important role to play.

Looking at the model of the stool of sustainability (Figure 1, page 11) and applying it to sustainability of food supplies:

- Environmental sustainability encompasses producing food without undue degradation of the soil, or overuse of water resources, or unacceptable levels of pollution, or degradation of ecosystems and habitats, yet at the same time meets the likely increase in food demand.
- Sociocultural sustainability allows rural communities to retain their traditions, cohesiveness and cultural values while producing sufficient food for their local areas without undue reliance on imports.

- Economic sustainability requires an agricultural system to provide an acceptable economic return for those employed in the production of food, yet at the same time supplying sufficient food to support a country's non-farming population.

Figure 4 applies the concept of the sustainability quadrant to the issue of food supplies.

Futurity	Environment
Conserves scarce resources, such as water and energy, yet provides secure supplies for the future and is proofed against climate change.	Uses methods that are eco-friendly and conserves and enhances biodiversity and landscape amenity for long-term security.
Public participation	**Equity and social justice**
Involves local people such as smallholders and local systems in developing and maintaining the economic vitality of rural areas and empowering them to make decisions.	Provides food security for all avoiding the extremes of globesity and famine and provides good quality food efficiently and at affordable prices.

Figure 4 The food sustainability quadrant

Many of these aims are lofty ideals that are, in reality, difficult to achieve as a result of conflicting views of the players involved in the food supply chain. Moreover, sustainability has to operate from a micro scale of a single farm unit to a macro-global scale, which again imposes tensions. There are many debates as to how best to achieve this — hence the series of World Food Summits after the 2008 crisis.

Figure 5 summarises the basic criteria for food sustainability across the food supply chain.

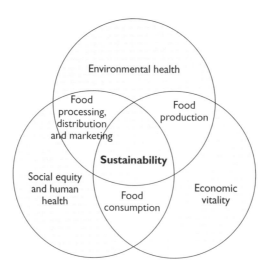

Figure 5 Food supply chain sustainability

1.2 What is the global pattern of food consumption?

Something is badly wrong with the way the world feeds itself. In rich countries many foods are cheaper than ever before, yet premature deaths from diet-related, obesity-induced diseases are soaring. At a farm level, soil, water and biodiversity are under pressure as never before from intensive farming or farming beyond the margin. In the poorest countries the situation is even worse. In sub-Saharan Africa, malnutrition and hunger are rife, and many peasant farmers often cannot compete with the flood of subsidised food from over-production on intensive farms in rich areas. As countries develop economically, especially in Latin America and Asia, their people adopt Western-style diets (the globalisation of food tastes), which results in diet-induced diseases such as diabetes.

Sustainability will only be achieved if relationships between governments, food companies, farmers and consumers can be transformed.

Under-nutrition

Food consumption can most easily be measured by daily calorie intake. Although the Food and Agriculture Organization (FAO) has calculated that enough food is produced to feed everyone in the world, in 2010 approaching 950 million people do not get sufficient calories to lead active and healthy lives. They are consuming fewer than 2000 calories a day, with too little protein and energy to sustain a healthy weight (**under-nourishment**), and suffer from deficiencies such as a lack of vitamins in their diet that leave them vulnerable to disease.

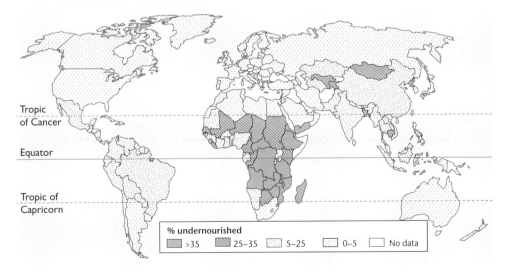

Figure 6 Global distribution of under-nourishment, 2003–05

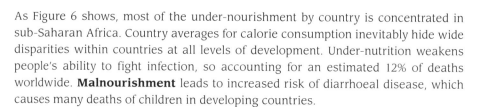

As Figure 6 shows, most of the under-nourishment by country is concentrated in sub-Saharan Africa. Country averages for calorie consumption inevitably hide wide disparities within countries at all levels of development. Under-nutrition weakens people's ability to fight infection, so accounting for an estimated 12% of deaths worldwide. **Malnourishment** leads to increased risk of diarrhoeal disease, which causes many deaths of children in developing countries.

Chronic under-nutrition is therefore not a consequence of overall scarcity and shortages of food but of unequal access to land, technology, employment and educational opportunities. This is combined with the range of socioeconomic and environmental factors shown in Table 3, which exacerbate food insecurity. For example in Ethiopia, an LDC, population densities are high, there is marked climate variability, including many seasons of rain failure, and widespread poverty, which means that families cannot buy food in times of shortage. Malnourishment has increased recently almost to famine level because of the rising cost of staples.

Table 3 Causes of under-nutrition

	Direct causes	Root causes
Economic	Poverty, land tenure and reform — landlessness, food supply from impacted production, food herding, poor infrastructure, storage, inappropriate aid	Unfair trade — trade restrictions, debt repayment (SAPs)
Social	Population growth, poor health and reduced labour (especially as a result of HIV), deliberate food destruction in wars, gender inequality	Military and civil conflicts, war and corruption, subsequent displacement of people (refugees), urbanisation (as people lose ability to grow their own food)
Environmental	Natural disasters of drought, desertification, flooding, pests, overcropping and overgrazing, development of biofuels and cash cropping	Short-term climate change, demands of green energies

Under-nutrition has a disproportionate impact on the young, the elderly and the poor, who suffer the most deaths. Although the overall production of cereals is globally sufficient — largely because productivity has improved substantially in South America and Asia (Green Revolution) — in sub-Saharan African countries such as Zimbabwe or Sudan, where the need is greatest, production has actually declined. Figure 7 shows the complex causes of food insecurity in Sudan.

Food is redistributed around the world via trade and aid, but both can be problematical and insufficient to solve the problem of under-nutrition in the long term.

(a)

Long-term factors leading to food insecurity			
Physical	**Social**	**Agricultural**	**Economic/political**
• Long-term decline in rainfall in southern Sudan • Increased rainfall variability • Increased use of marginal land leading to degradation • Flooding	• High population growth (3%) linked to use of marginal land (overgrazing, erosion) • High female illiteracy rates (65%) • Poor infant health • Increased threat of HIV/AIDS	• Highly variable per capita food production • Static or falling crop yields • Low and falling fertiliser use • No food surplus for use in crisis	• High dependency on farming (70% of labour force; 37% of GDP) • Dependency on food imports • Limited access to markets to buy food or infrastructure to distribute it • Debt and debt repayments limit social and economic spending • High military spending

Short-term factors leading to famine

Drought in southern Sudan. Any surpluses quickly used up

Both reduce food availability in Sudan and inflate food prices

Migration from Darfur towards areas already under food stress

Situation compounded by:
• Lack of political will by government
• Slow donor response
• Limited access to famine areas
• Regional food shortages

(b)

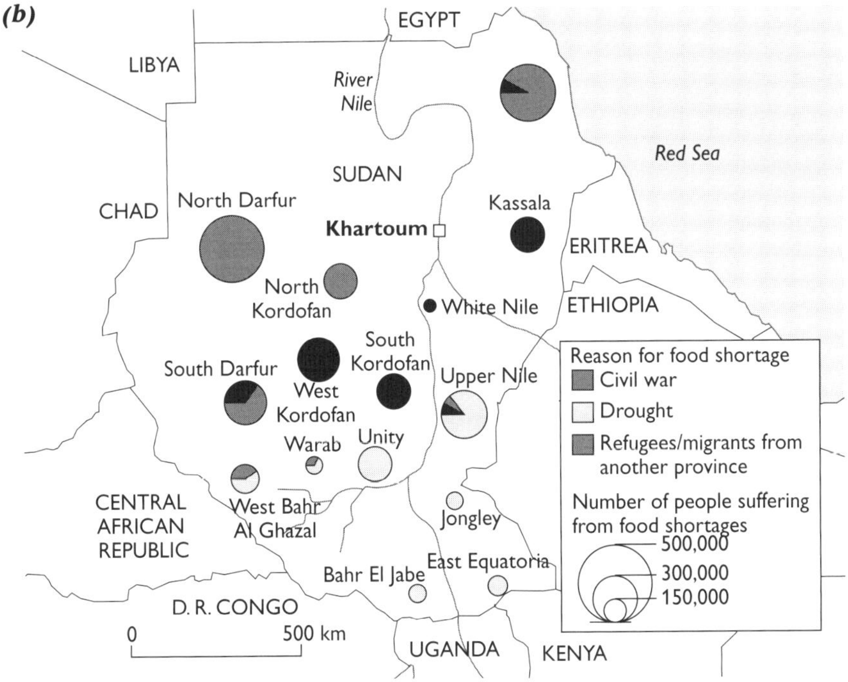

Figure 7 Complex causes of food insecurity in Sudan:
(a) causes of famine (b) pattern of food shortages

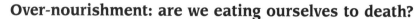

Over-nourishment: are we eating ourselves to death?

Overweight and obesity are defined as abnormal or excessive fat accumulation that may impair health. Body mass index (BMI) is an index of weight for height (weight in kilos/square of height in metres), which is used as a numerical measure for comparison. According to World Health Organization (WHO) definitions:

- overweight occurs when BMI is equal to more than 25
- obesity occurs when BMI is equal to more than 30.

However, in some Asian and middle eastern countries such as Mongolia, Thailand and Oman, recent research suggests that a BMI of 22 combined with a poor diet, high in sugar and fat, may lead to ill health.

Globally, 1.6 billion adults are overweight (25% of the world) and over 400 million adults are classified as obese. Of even greater concern are the WHO projections of around 40% of population overweight and 750 million obese by 2015.

A common sight in many countries are obese children (some 20 million globally in 2010), particularly in the USA where 35% of children are clinically obese and increasingly in China where the transition to a Western-style diet has generated some well fed cuddly 'Little Emperors'. The WHO is currently developing an international growth reference standard for school-age children and adolescents in order to increase the reliability of the statistics for childhood obesity. So widespread is obesity as a global problem that the term **globesity** has been coined to describe it.

The problem is caused by:

- the global shift to a Western-style diet with an increased intake of energy-dense foods — often ready meals and fast food high in fat and sugars, which leads to increased inputs
- a trend towards decreased physical activity due to the increasingly sedentary nature of many forms of employment and the changing modes of transportation (going everywhere by car), as well as a generation of children who are wedded to computer games and social networking websites rather than outdoor team games

There are major health consequences of over-nutrition, in particular the huge rise in chronic conditions such as cardiovascular disease, diabetes, musculoskeletal disorders such as osteoarthritis and possibly some cancers (for example, colon), all leading to premature death or disability and reduced life expectancy. As a result, many low- and low–middle income countries are now facing a 'double whammy' for their embryonic health systems of dealing with both chronic and infectious diseases.

Another issue is that as the diet transition occurs, there is more dependency on globalised trade in foodstuffs with products traded around the world — leading to the high carbon footprint from these food miles so apparent in surveys of supermarket TNCs (see Figure 12 on page 34).

Figure 8 shows the prevalence of male and female obesity.

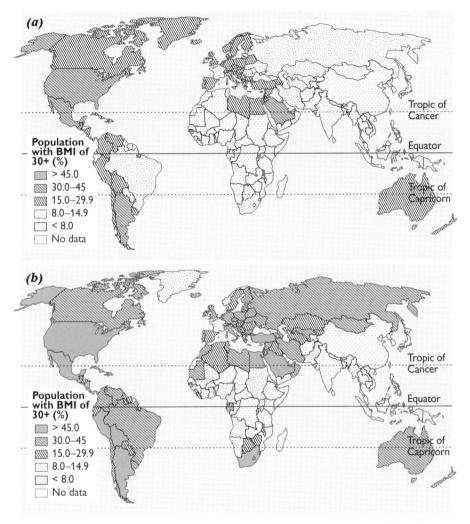

Figure 8 Prevalence of obesity aged 30+, 2005: (a) males, (b) females

What Figure 8 shows is that there is not a straightforward correlation between levels of affluence and obesity. Although the percentages are high in developed countries, in certain areas such as the Caribbean and South Pacific levels of obesity are high for sociocultural reasons, especially among females (diet and the preference for traditionally built women). A further point to make is that the maps are for 2005, but by 2015 the pattern will have changed dramatically because of diet transition in Asia.

As a drive towards sustainability and human wellbeing, it is vital that measures must be taken to resolve both problems of under- and over-nutrition. For example, obesity could be cut by initiatives such as reducing portion sizes in restaurants; introducing and promoting healthy nutritional choices (school meals); informing people and reducing the fat, sugar and salt content of processed foods (supermarket labelling); and promoting healthy exercise and nutritious food choices (free fruit for children in deprived areas).

Figure 9 summarises the range of nutrition discussed in Section 1.2.

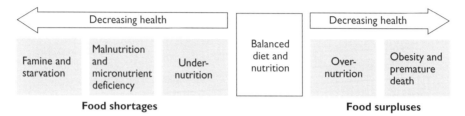

Figure 9 The nutrition spectrum

1.3 What factors promote or hinder food production?

A number of factors, both physical and human, combine to explain why food supply varies spatially. Some areas, such as the Prairies or the Russian steppes, have physical advantages such as fertile, deep-black chernozem soils, which make them key suppliers of food. Other areas, such as Java, have high-quality volcanic soils and support a dense rural population. In all cases, even areas with high output can be disrupted by the impact of El Niño/La Niña or extreme weather conditions. Figure 10 shows how a number of factors can influence food production systems.

The **physical** environment tends to operate as a constraint on agriculture, in that it is the land that provides all the basic essentials to support agro-ecosystems: heat, sunshine, water and soil.

Climate is generally regarded as the most significant control as it exerts direct influence on the growth and survival of plants and animals in terms of temperature, rainfall patterns and growing season length. It also has an indirect influence, for example on soil formation. Soils can display a variety of characteristics such as depth, degree of acidity, moisture-holding capacity, texture (workability) and susceptibility to erosion. Certain crops vary in terms of the type of soil they require to flourish. The relief of the land can affect agricultural activities in a number of ways. For crops such as orchards, aspect (sunshine hours), altitude (late frosts) and slope (cold air drainage) are all significant.

Of increasing importance are **human** factors, which include socioeconomic, political and technological factors. As Figure 10 shows, there are numerous factors that impact on levels of food production. Key economic factors include:
- Farm size and tenure can place limitations on the farmer when making land-use decisions. Although size variation is largely a function of land quality and land

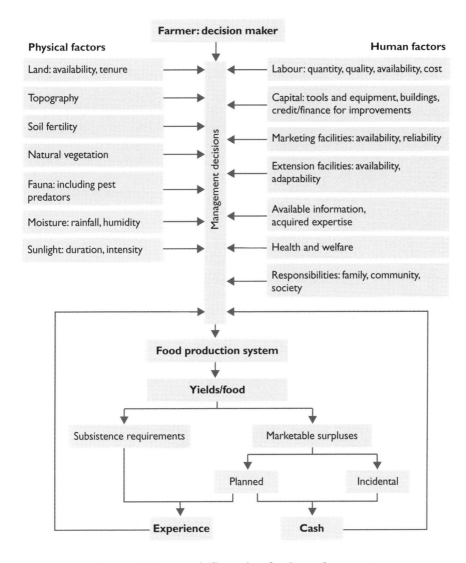

Figure 10 Factors influencing food supply systems

availability, population density, how the area was settled (for example, in the Prairies) and inheritance laws are also important. These factors play a major part on how intensively the land is farmed. If a 1 ha area has to be farmed intensively in order to support a family, or if the carrying capacity of the land is low and it is located in a marginal area, only a large farm would provide enough pasture or land to support a viable stock-rearing or arable enterprise. Large farms use extensive methods but are frequently dependent on highly technological methods as they can benefit from economies of scale.

- Land tenure, defined as 'the economic, political and legal aspect of ownership and management of agricultural land', can have far-reaching impacts on decisions about agricultural practice. The principal forms are: owner-occupation (common in capitalist countries); various forms of tenancy (including share cropping); and communal ownership (common in tribal Africa or in association with Communist-style economies).

- Capital usually refers to fixed capital (i.e. the fabric of the estate) and working capital (livestock, machinery, seeds, fertilisers). Finance usually has to be raised through banks and other private institutions. In developed countries most agriculture is capital intensive, with the capacity to modify aspects of the physical environment to achieve maximum yields (for example, fertilisers and pesticides). Although successful agribusinesses can raise capital, a lack of capital is a factor that limits productivity on subsistence peasant farms. It is here that aid from non-governmental organisations (NGOs) such as Farm Africa and ActionAid, the formation of trading cooperatives and loans from institutions such as Grameen Bank can play a vital role in securing food supplies.

- Labour has traditionally been the most important factor of production. Many subsistence farms and certain types of cash cropping agribusinesses are labour intensive. Fruit and vegetable farms in particular are dependent on abundant supplies of labour (usually migrants) for harvesting (often year-round) in order to sustain their production. Exploitation of the workforce in order to achieve profitability has no place when considering sustainability.

- Both transport and markets were traditionally important as limiting factors to the farmer's choice of crops. However, a series of developments in transport and food production technology has diminished their control as limiting factors. Agribusinesses increasingly control the growth, processing, distribution, marketing and retailing of food worldwide. Global markets have developed for fruit and vegetables as TNCs have set up subsidiaries in developing countries for the high-value market of the developed world — using air freight and refrigeration technology. It is here that issues of 'food miles madness' emerge. There are costs in terms of diverting land in food-deficit countries from producing food for local consumption into cash crops for export. There are also costs in terms of air pollution and carbon dioxide emissions in transport around the globe. Sometimes this means that imported foods flourish at the expense of home-grown farming as supermarkets promote perfectly sized foreign produce. All these factors make the type of production almost totally unsustainable, yet there are positives, such as the way agribusinesses provide secure jobs for many workers as not all TNCs are totally exploitive.

- Market prices are a key issue at many scales. The problem of price instability and exploitation is a major one for independent commercial farmers and estates in the developing world as they struggle to get a fair price for their commodities such as tea or bananas. A complex web of subsidies, quotas and tariffs and world trade conditions has produced a crisis currently in the sugar production industry.

- Government intervention has been identified as a major and increasing influence on present-day agricultural systems. A study of the EU agriculture policy and its impact on UK agriculture reveals three main phases of development.

Phase 1 1947–72: revitalise farm production and provide cheap food as part of postwar recovery by promoting self-sufficiency to improve food security via price support and subsidies.

Phase 2 1973–85: provide security to rural areas by supporting farm and other rural activities, especially in less favoured areas.

Phase 3 1985 onwards: control production to prevent surpluses and reduce environmental damage by a series of production controls (quotas, set aside, grants for farm diversification) and environmental initiatives such as the creation of environmentally sensitive areas (ESAs) and nitrate sensitive areas (NSAs), together with environmental stewardship schemes (MacSharry reforms post 1990).

Phases 1 and 2 were concerned with economic and social sustainability whereas Phase 3 favours environmental sustainability, showing how the various facets of sustainability can conflict.

Government policies are affected by a range of concerns such as prices of farm products in supermarkets and financial returns to farmers. Currently there is a war of attrition between farmers, supermarkets, government and consumers to try to balance these two conflicting concerns. Many would regard the move to ethical consumption — supporting fair trade products and high-quality 'welfare' food — as improving the sustainability of food supplies.

- Technological innovation has a huge contribution to make towards food security as increased inputs raise outputs. The following section asks how food production can be increased sustainably; many of the 'techno fixes' such as hydroponics, fish farming or the green or gene revolution come with costs as well as benefits. High technology is not necessarily the answer, and for many developing areas the role of intermediate technology is vital in increasing food production. The crucial point is, to be sustainable, the technology must be appropriate.

1.4 Can food production be increased sustainably?

There are two ways in which food production can be increased: by **intensification**, in which higher yields are demanded from existing farm lands; and by **extensification**, in which new areas are farmed, either in more marginal lands or in areas in which severe limitations have to be overcome by technological innovation. Between 1995 and 2010, 80% of the total increase in food production came from intensification. This trend is likely to be even more marked in the future, although to maintain global food supplies a 5% expansion in land used for agriculture is needed, especially in developing countries where it is more available (less urban sprawl).

Intensification is traditionally achieved by raising levels of inputs using a number of strategies such as increased mechanisation, plant and livestock breeding, irrigation and the application of agrochemicals (fertilisers, pesticides and drugs such as antibiotics and growth hormones).

Table 4 Farming inputs and outputs: environmental impacts of high-tech farming

Input	Purpose	Outputs — environmental impacts
Machinery, e.g. tractors and combine harvesters, and air transport for perishables	To replace human or animal labour, and increase efficiency	Pollution: increased fossil fuel use for transport and refrigeration. Increased food miles and packaging. Soil compaction and erosion. Loss of biodiversity in wetlands, hedgerows, forests. Trawling impacts on marine ecosystems and non-target species, e.g. dolphin, albatross
Chemical fertilisers, especially nitrogen products	To increase yield by providing high levels of plant nutrients	Pollution: eutrophication of water by agricultural runoff, loss of biodiversity, e.g. sugar cane in Caribbean
Pesticides	To remove insects and other pests, which could reduce yields	Toxic chemicals, especially DDT, entering the food chain and causing damage to organisms that were not the intended 'victim' of the chemical
Herbicides and fungicides	To remove weeds, which take up space and use up nutrients, and to reduce fungal diseases that reduce yields	
Antibiotics	To increase resistance to disease in livestock and fish and to increase yield	Antibiotic-resistant bacteria and the danger of epidemic outbreaks; fears for human health due to consuming this meat
Animal/fish feed	To increase the density of animals/fish kept in a given area e.g. using battery hens, feedlots and aquaculture	Increased food crops to be used to feed livestock, and therefore increased pressure to clear areas (e.g. forests) to produce more crops especially soya bean. Waste products from animal slurry are highly toxic. Methane from livestock is a potent global warming gas. Intensively reared cattle are fed diets rich in protein and energy. For every acre of feedlot in the UK, two more are farmed overseas to meet its needs

As the following case studies will show, there are environmental and economic costs involved that can outweigh the benefits of improved food security and challenge the achievement of all facets of sustainability, especially as fertilisers have been the main driver of rising food production and have led to massive eutrophication.

Extensification has its drawbacks too: the use of high technology can be costly and sometimes inappropriate and most of the areas with good farming potential have already been used. Marginal lands such as semi-arid areas may be fragile and any farming developments can lead to soil and land degradation and the onset of desertification. Moreover, many of the other potential areas for expansion such as the Atlantic and Amazon rainforests of Brazil (under pressure from soya bean production for biofuels) are of outstanding ecological value (designated hotspots or eco-regions) and therefore need protection from human activities and designation as some form of reserve, with conservation status for their threatened species.

The Green Revolution

The **Green Revolution** occurred in the second half of the twentieth century and contributed to much of the increase in agricultural production since the 1960s, especially in countries in Asia and South America. In 1961, 1.84 billion tonnes of food were produced; in 2007, 4.38 billion tonnes. The Green Revolution involved breeding high-yield varieties (HYV) of wheat, rice, maize, sorghum and millet to increase crop yields and stabilise food supplies. It was supported by a package of agricultural improvements including mechanisation, fertilisers, pesticides and herbicides, often using irrigation in order to facilitate double cropping in a monsoon environment with marked seasonality of rainfall. Fertilisers are considered by researchers to be the key component of the increases in cereal yield. Figure 11 assesses the costs and benefits of the Green Revolution.

(+) The world is better off because there is far more food available from bigger harvests than 20 years ago

(−) Bigger granaries storing more grain are not enough; hungry people need to be eating more; food in storage is not being distributed to the needy

(+) New farming methods with irrigation can bring all-year-round employment; no longer do workers have to be laid off in the dry season; double cropping is common

(−) Often, agricultural profits are invested in tractors, which reduces employment

(+) Significant and populous countries such as India, Indonesia and Thailand have become self-sufficient in basic foodstuffs; they are no longer dependent on North American and European food aid

(−) But implementing Green Revolution policies has increased dependency on imported seeds, fertilisers, pesticides and farm machinery

(+) With bigger harvests, farmers earn more and the price of food stays constant or even becomes cheaper in the marketplace

(−) For the small farmers who cannot afford the Green Revolution seeds and other technology, the lower prices for their harvests means real hardship — often they have to sell off to the big landowners, bringing an increasing gap between rich and poor

(+) Former food importers such as India and Thailand now export grain, earning useful foreign exchange; initially the taste of HYV rice products was a problem

(−) Imports of seeds, petrochemical fertilisers and fuel for machinery all cost valuable foreign exchange, and give control to agrochemical multinationals

(+) The environmental impact of impoverished rural people — who cut down trees — is lessened by urban migration

(−) New farming methods can bring an increase in water-borne diseases (with irrigation), the development of 'super-pests' (resistant to insecticides), and desertification (through the salinisation of waterlogged fields); increased fertiliser use leads to eutrophication; biodiversity is lost as native breeds are replaced by HYVs

---

Assessment

Increasing food production does not stem hunger on its own. It is possible to have both more food and more hunger. 'If the poor don't have money to buy food,' a World Bank report said, 'increased production won't help them.' Nevertheless, increasing harvests by 30% is the equivalent of discovering 30% more farming land — and as the population is increasing, that is a welcome relief.

Figure 11 Benefits and costs of the Green Revolution

There are many features that question the sustainability of the Green Revolution, such as the impact of agrochemicals on the environment, increased soil salination resulting from irrigation, the dispossession of many small farmers, and increased unemployment resulting from mechanisation which worsened the lives of many rural families. On the other hand, the gains in cereal productivity have been dramatic, with output doubling between 1960 and 1990, thus increasing food security, with famines now a rarity in countries such as India, the Philippines and Thailand.

Recent concerns have been raised about the nutritional value of the high HYVs, especially as they have displaced many local fruit and vegetables that traditionally supplied nutrients in people's diets.

There is now talk of a second Green Revolution in Africa. The Alliance for a Green Revolution in Africa (AGRA), an organisation funded by several US donors such as the Rockefeller and Gates Foundations, has introduced specially bred crops, including Nerica rice, which has a short growing cycle, resists weeds and therefore gives double yields. Current research is focused on developing drought-resistant varieties of cassava, maize and millet, which will help subsistence farmers cope with climate change. The Green Revolution will only work in Africa if the varieties of crop used will grow in difficult conditions (drought and low-fertility soil) and do not require large quantities of costly inputs of irrigation water, fertilisers and pesticides. Equally, there is a need to develop supporting technologies to ensure farmers can store and market their surpluses.

Many of the most favoured areas of Africa have now been taken over by TNCs to grow cash crops, so peasant farmers have been pushed to less favourable areas. This second Green Revolution has several more obvious features of sustainability, especially as the farmers are trained to use methods such as **integrated pest management (IPM)**. This is a strategy that uses an array of complementary methods, natural predators, pest-resistant varieties and biological controls (disruption of mating). Pesticides are used only as a last resort, making it a sustainable strategy as it is 'friendly' to the environment and people's health. There is also a move to support the farmers with micro-credit facilities, and schemes to share new ideas for the farmers to work together — a feature of sociocultural sustainability.

The gene revolution

Although selective plant and animal breeding has been undertaken for many years, the gene revolution is a product of modern biotechnology that involves modifying and therefore altering the DNA of crops (known as **genetic modification**, or GM) by taking some of the DNA from one species and adding it to the DNA of another; for example, adding the genes of a herbicide-resistant weed to a wheat seed so it is not harmed by herbicides.

Inevitably there are many controversies concerning this revolutionary development. Initially the development has been largely in the hands of TNCs such as Monsanto. The main crops have been herbicide-tolerant soya bean and maize, which will produce higher yields because of the absence of weeds that compete for nutrients and water, and Bt cotton (cotton that is modified to combat insects using its own bacteria).

Table 5 summarises the arguments against first-generation GM crops. There are clearly doubts about the environmental, sociocultural and economic sustainability of such technology.

Table 5 Arguments against first-generation GM crops

Many GM crops are designed to be immune to one, very strong, herbicide. If this is sprayed on the crop, all weeds are destroyed. This could virtually eliminate biodiversity. The GM crops may also hybridise with natural species, some of which may be weeds. In the hybridisation, they may acquire the resistance to herbicides from the GM crops. Their spread would then be uncontrollable.
Research and development is only really possible by a few truly giant multinational corporations as the funding required is vast. If they become the standard crops, all farmers will have to buy seeds from just a small number of firms or possibly just one. In order to control weeds and pests, farmers would be forced to buy pesticides and herbicides that cannot harm the crop and which are manufactured by that one company. This would place the future of the world's food supply into the hands of a small number of corporate bosses. They would exert massive power over almost all governments of the world.
GM crops produce good yields but the seeds are infertile. Farmers would be forced to buy new seeds each year. In the past, farmers saved seeds from the harvest to sow the next year.
One of the main reasons for developing GM crops is to prevent food shortages. Many groups argue that there is no shortage; the problem is the distribution of food throughout the world. This argument may well have been true in the past, but other evidence suggests that this argument is no longer valid.
Those who generally favour organic food are outright against GM crops. Others claim that they would be harmful to human health, but no evidence to support this has been proved so far. There are groups that claim it is going against nature and playing God with the Earth's resources. Such crops are often described as 'Frankenstein crops'. As yet, none of these groups has produced hard evidence. Many rational people wonder if their concerns should be listened to. For example, if problems are found in the future, but we are more or less committed to GM crops, how can the clock be turned back?

However, as GM technology has developed, second-generation crops are beginning to answer many of the initial concerns. New varieties are being bred to overcome many of the issues that significantly reduce crop yields or cost farmers money:

- They will need less fertiliser by using nitrogen more efficiently.
- They will withstand environmental stresses such as drought, frosts and saline soils, and therefore could be used on marginal lands.
- They will be disease-resistant and pest-resistant.
- They can deliver enhanced nutritional value by adding nutrients that are lacking in the diets of many people.
- The toxins can be removed so that cotton seeds are used as a food protein supplement.
- They can be modified to produce vaccines.

Moreover, in addition to the leading growers of first-generation GM such as Brazil, Argentina and USA, both China and India have become leaders in genetic engineering and have therefore developed strong biotech industries. This may ultimately lead to an improved range of crops including other staples such as cassava, potatoes, millet and wheat, as well as fruit such as tomatoes, and also an expansion of research to look at how GM can benefit small-scale, largely subsistence farmers and the world's poorest peoples as opposed to commercial farmers in rich countries.

The Blue Revolution

Although both freshwater and marine resources have always been a source of food for people, there are huge issues about overfishing and whether oceans, rivers and lakes are a sustainable option in the drive to feed the world's population. The depletion of fish stocks all around the world, such as cod and herring in the North Sea, has led to the **Blue Revolution** of aquaculture whereby fish are farmed in controlled environments. Aquaculture is growing at a rate of nearly 10% per year and now accounts for one-third of all 'catches' of fish, crustaceans and molluscs by weight, especially in Asia and the Pacific region, which accounts for 90% of the production quantity and 80% of its value, with China and its ubiquitous fish ponds and farms being the world leader.

On the face of it, the concept seems to be a sustainable one, but there are environmental issues such as pollution (from feedstuffs), disease, food safety and public health. There is also the fact that the farmed fish such as salmon need to be fed on young fish and krill, which puts the sustainability of the whole food chain at risk. Therefore, aquaculture is a definite plus for food security, but concerns exist over both the futurity and the environmental elements of the sustainability quadrant. So much depends on the scale and intensity of the production, with great concerns expressed over large, commercially intensive prawn and salmon farms.

Technological advances

If technology is to enhance sustainability, there is a need to link the fruits of scientific and technological research and development to the frontline of farmers — 'translating the field lab into farming' (J Smith). Technology may provide theoretical answers to problems of saving water and energy in agricultural production, and protect the fertility of the land, but it requires farmers to be trained on how to use it and apply it to maximum effect in local conditions. Moreover, agro-technology by itself is not enough, as it must be accompanied by strategies to consider local socioeconomic and environmental needs. The key is to combine science and technological research with traditional knowledge in order to provide locally appropriate solutions, and for this the technology may be of any kind. The images of the Russian tractors donated to African countries in the 1960s, rusting even today in villages for the lack of spare parts, still prey on the minds of those opposed to technology.

High technology

The technological battle to raise agricultural productivity while reducing the environmental impact of agrochemicals and irrigation is taking place across a wide front. Although genetic engineering of crops (biotechnology) has received most publicity, there are many other promising developments.

Precision farming, sometimes known as **smart farming**, uses information technology to monitor crop growth and harvesting and guides the application of agricultural chemicals and water, i.e. more efficient fertilisation and more crop for your drop!

Satellite navigation is used to guide tractors to avoid overlapping of seed planting within an accuracy of 2 cm, thus saving fuel and seed supplies. Precision farming also allows images from Earth observation satellites to be used to monitor individual fields, and adjusts agricultural inputs such as fertiliser by micro dosing according to crop and soil conditions. The radiation from the crops is reflected into space so that satellite images will reveal crop health, levels of moisture, nutrients, soil properties and likely yields. Access to satellite technology is potentially available to all countries in the world — at a price. Interestingly, the equipment on the tractor itself only adds around 10% to the cost of a new, fully equipped tractor, but clearly is most beneficial to larger farms that can use it more intensively.

PNT technology is used to release fertiliser in a controlled way in the form of an amine (a nitrogen compound), which reduces wastage and pollution through leaching into soil. **CaT technology** helps plants absorb calcium more efficiently, so alleviating the environmental impact of drought. A third technology, called **Alethea**, aims to protect plants against stress by strengthening their cell walls. Initially, it is being used in high-value horticultural markets in countries such as the Netherlands, but in time it will be used on cereals and may be applicable worldwide.

Similar high-tech advances have been made in **irrigation technology**, in particular using drip irrigation which delivers water at or near the root zone at the most appropriate time, minimising evaporation and runoff. This technology is widespread in all commercial farms.

Infrared technology is used to identify problems with soils. Infrared spectroscopy is being used to halt soil degradation in Kenya. In the Sahel infrared technology is used as one component in the UN Millennium Villages project, which aims to halt the southward creep of desertification by planting trees.

Hydroponics and **aeroponics** can provide high outputs per unit area of clean, disease-free food, as all the crops are grown in the controlled environment of large greenhouses, with targeted nutrients and water to ensure maximum plant growth. The disadvantage, unlike the above examples, is the high demand for energy inputs. The latest trend is to develop **vertical farms**, which stack hydroponic areas and lead to space saving (see WJEC paper Unit 4, January 2010).

In conclusion, certain types of high technology can be sustainable in that they make production more efficient by maintaining yields yet using inputs more effectively. Appropriately used, they can make a major contribution to increasing food supplies by more efficient production.

Intermediate technology

Intermediate technology also has a major role to play as it uses low-tech ideas (cheap to build and maintain, and applicable to a local situation as the people can be trained to use it, usually by NGO field workers). There are numerous examples of these technologies, all of which can be considered sustainable as they help to conserve resources and involve local communities in the future of their food security. Examples include:

- agro-forestry — growing crops beneath trees
- combining organic, labour-intensive farming with compost-making, vermi-composting and **permaculture**. Permaculture (as practised in Cuba) is both high yielding and sustainable as it is based on, and takes advantage of, natural ecological processes and relies on diverse farming where processes are integrated
- using a range of simple machinery to irrigate the land such as pumpkin tanks to store water during the dry season and treadle pumps to deliver it to the land
- technologies for drudgery reduction such as fuel-efficient cookers that cut down on fuel-wood gathering, thereby allowing women more time for agricultural work
- new technologies of animal husbandry such as zero-grazing technology
- using innovative means of pest control such as the push–pull system of pest management — plant species that push away pests or pull them into 'trap' crops. For example, enclose maize fields in a border of Napier grass, which is more attractive to moths than the maize (pull) and plant the forage legume silver leaf in the maize, which fixes nitrogen and releases chemicals that repel the moths (push).

Exam hint

Research key websites such as **www.practicalaction.org** or **www.other90.coop-erhewitt.org** for full details of the huge range of intermediate technologies being developed.

Other developments: sustainable food chains

Via Campesina is a global international peasant movement that favours food sovereignty over food security whereby food production and consumption is organised according to the needs of local communities. A fundamental tenet of the organisation is that peasant or family farm agriculture is based on sustainable production using local resources in harmony with local traditions and culture — an agro-ecological approach.

Table 6 summarises the main features of such integrated farming systems, which involve using many varieties of crops, local energy and labour resources.

Table 6 Integrated farming systems

Principle	Characteristics
Crop rotation	Promotion of soil structure and fertility and reduction of demand for agrochemicals; minimum of four different crops in rotation is recommended
Minimum soil cultivation	Provides agronomic and environmental benefits, e.g. reduced soil erosion and nitrogen volatilisation. Use of mechanical tools for weed control
Disease-resistant cultivars	Enables reduced use of inputs
Modifications to sowing times	For example, later sowing times reduce pests and outbreaks of disease
Targeted application of nutrients	Saves costs by reducing the amount of chemical applied and provides environmental benefits, e.g. by reducing chemical contamination of groundwater
Rational use of pesticides	This can include avoidance of prophylactic spraying through crop monitoring and using thresholds to determine the most appropriate timing of application
Management of field margins	Creates habitats for predators
Tillage systems	To favour natural control of pests, improvement of soil structure and to reduce demand for external nitrogen
Cropping sequences	Modifications to increase crop diversity
Promotion of biodiversity	Provides ecological benefits and promotion of beneficial predators. Between 3% and 5% of total cropping area is usually recommended as non-agricultural vegetation

The aim of an integrated farming system is for the produce to be consumed locally, cutting down on energy loss and fuel bills, again increasing sustainability. Farmers' markets and farm shops provide outlets for local produce.

Urban farms are springing up around and within cities, working on the same principle of linking the producer and consumer together, again cutting down on food miles. Figure 12 shows the food kilometres of a UK supermarket. Think how integrated farming systems are so much more sustainable.

Another development that enhances the sustainability of food production is **ethical consuming** whereby individual purchasers make the connection between the products in the stores, where they come from, under which conditions they have been produced and how they have been packaged. At the heart of this movement is **fair trade** whereby producers in developing countries form cooperatives to market their products for which they are paid a fair price, thus not exploiting the poor farmers. TNCs and the supermarkets commit to promoting fair trade products, and consumers agree to purchasing them at a higher price. Although the sale of fair trade goods has increased by 60% in the last 5 years, the sector accounts for only a small percentage of total trade and many argue it is **fairer trade** — which involves the reformation of World Trade Organization rules for free trade — that is more fundamental to improving food security. The dumping of subsidised farm produce from developed countries inhibits the growth of farming in developing nations.

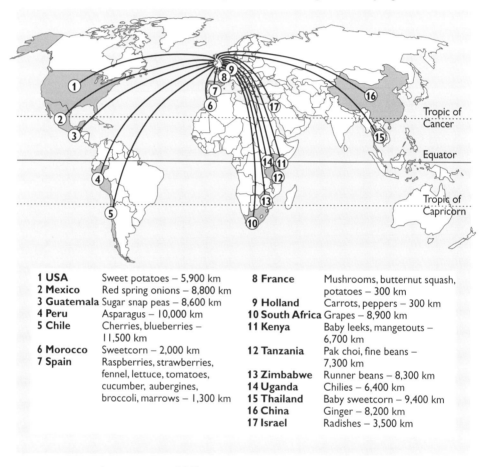

1 USA	Sweet potatoes – 5,900 km	**8 France**	Mushrooms, butternut squash, potatoes – 300 km
2 Mexico	Red spring onions – 8,800 km	**9 Holland**	Carrots, peppers – 300 km
3 Guatemala	Sugar snap peas – 8,600 km	**10 South Africa**	Grapes – 8,900 km
4 Peru	Asparagus – 10,000 km	**11 Kenya**	Baby leeks, mangetouts – 6,700 km
5 Chile	Cherries, blueberries – 11,500 km	**12 Tanzania**	Pak choi, fine beans – 7,300 km
6 Morocco	Sweetcorn – 2,000 km	**13 Zimbabwe**	Runner beans – 8,300 km
7 Spain	Raspberries, strawberries, fennel, lettuce, tomatoes, cucumber, aubergines, broccoli, marrows – 1,300 km	**14 Uganda**	Chilies – 6,400 km
		15 Thailand	Baby sweetcorn – 9,400 km
		16 China	Ginger – 8,200 km
		17 Israel	Radishes – 3,500 km

Figure 12 Food kilometres for UK supermarket sales

1.5 Can a sustainable food supply be maintained in the future?

At a local scale, the issue of whether a sustainable food supply can be maintained in the future depends on the rate of population change; given the encouraging developments in technology, some areas are increasingly able to provide food for their population and even export a surplus. There are, however, problem areas where — as a result of environmental factors such as the impact of short-term climate change leading to increasing extreme weather and concerns about water security, as well as war-induced famine and general poverty — food aid is required to feed people. This is the case in Darfur in Sudan and in northern Kenya, where pastoralists and agriculturalists are fighting for the limited water supply.

On a global scale, although the food crisis of 2008–09 is easing, there is no overall shortage of food as more than enough is produced. It is the systems of distribution, such as trade and aid, that need reforming. Therefore, if we look to 2050 and the daunting target of feeding an estimated 9.1 billion people against the background of the fear of reaching the tipping point in climate change, there are real concerns, especially as currently 1 billion people suffer from malnourishment or under-nourishment. Remember, however, that climate change is not all bad as considerable areas of the world, such as central Africa, are expected to show potential increases in output.

However, in the words of the head of Oxfam, 'It will require an increase in global political will compared with what we've seen. It's possible, but there has to be a recognition that this is about people's right to food rather than the interests of very large corporations who are currently driving the International Agenda.' There are examples where countries such as Brazil have successfully reversed the tide of food insecurity through proactive programmes integrating agriculture and food.

'Overall it's the issue of food entitlement — it's more a question of access to food than overall production. The resources are there. It requires policy choices and appropriate aid and investment.' (FAO)

Figure 13 summarises the strategies needed to successfully tackle food security.

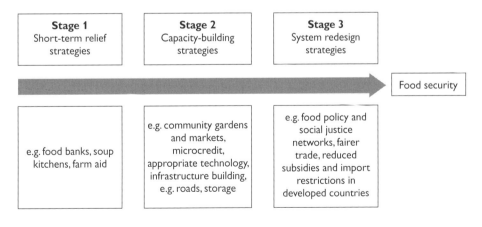

Figure 13 Food security strategies

Theme 2:
Sustainable water supply

2.1 The issues

Water, like energy and food, is a fundamental need. Political players would argue that all these basic needs are best provided by market mechanisms, but social players see access to available, affordable, safe supplies of clean water as a human right for all the world's peoples. Access to water and sanitation underpin all the Millennium Development Goals (MDGs). The problem is that experts predict there will be a **world water gap** between rising demands and dwindling supplies of what is essentially a **finite** resource, for which there is no substitute.

Global population growth (possibly an additional 3 billion people by 2035), economic development (especially in RICs and NICs) and rising standards of living all increase demand for water. If we follow a business as usual approach, there will be a 56% increase in demand on the 'global water pot' by 2025. Even if sustainable development strategies are adopted, an estimated 20% increase in demand is predicted. This may lead to conflicts between various users at all scales.

Water supplies are spread unevenly across the world. Two-thirds of the world's population lives in areas receiving only 25% of the world's annual rainfall. This leads to **physical scarcity** for many areas. It is a second potential source of conflict between countries and regions with a water deficit and those with surpluses, especially where they share a large river basin. These conflicts may be exacerbated by climate change, which may lead to greater future water scarcity.

A third underlying conflict arises from the **water availability gap** between the 'have nots' largely in developing nations (especially in sub-Saharan Africa) and 'haves' in the developed world. This gap is widening and reflects the development gap as many of these 'have nots' actually have enough physical supplies of water but suffer from **economic scarcity** in which the development of supplies is limited by a lack of capital and technology to pay for the exploitation. The lack of access to sufficient, safe (clean) and affordable water is termed **water insecurity**. The achievement of secure water supplies is vital to both economic development and human and ecosystem wellbeing.

There is also an imbalance in usage, with rich countries using up to ten times more water per head yet frequently paying much lower prices per cubic litre.

Human actions have led to fragmented management of supplies and to their widespread pollution.

Water scarcity, stress and vulnerability

Figure 14 summaries the global distribution of variability of water supplies.

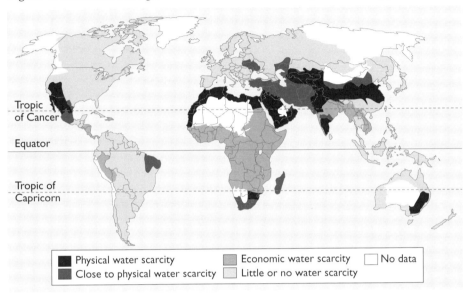

Figure 14 Global distribution of water scarcity, 2006

Developed countries in zones of secure water supplies generally have no problems.

- Scarcity occurs when the annual supply of water per person drops below 1,000 m³. Figure 14 shows the two types of water scarcity.
- Physical scarcity occurs when more than 75% of a country's or region's blue water flow (rivers and aquifers) is being used (25% of the world's population lives in such areas).
- Economic scarcity occurs when the development of adequate blue water sources is limited by a lack of capital and technology.
- Water stress occurs when there is less than 1,700 m³ of water per person per year. Those areas of water stress in the middle east, sub-Saharan Africa and southwest USA are likely to become areas of scarcity unless supplies are managed more sustainably.
- By 2025 it is estimated that nearly half of the world's population will be water vulnerable (less than 2,500 m³ of water per person per year) as supplies diminish and costs increase. Areas of concern include Mediterranean France and Spain and southeast England, where a variety of strategies will be needed to manage water shortages. By 2050 climate change may convert the vulnerability to stress (affecting an estimated 4 billion people) and even scarcity (a further 1.5 billion people). In 2010 the average UK availability of water is 2,500 m³ per person, much less than France or Italy. In southeast England — with a dense population of 425 people per km² and rising and an effective rainfall of only 266 mm per year — there is only 610 m³ of water per person, less than countries like Egypt.

Defining sustainable water supply: development and use

The concept of water sustainability is enshrined in the notion of water security. For all the world's nations, communities and peoples, clean, safe water should be available, accessible and affordable (the 3 As). The concept of the three-legged stool (Figure 15) and the water sustainability quadrant (Figure 16) will help you to consider what this actually means.

Figure 15 The stool of water sustainability

Futurity	Environment
Energy efficiency and economy of use (conservation) to manage demand yet at the same time ensuring security of supplies for the future.	Achieving high standards of environmental protection. Restoration of damaged water supplies.
Public participation	**Equity and social justice**
Involvement of communities. Decentralised decision making to ensure bottom-up appropriate technology solutions, e.g. from NGO WaterAid.	Equitable allocation between users to ensure secure supplies at affordable prices delivered by good governance and management.

Figure 16 The water sustainability quadrant

Environmental sustainability is a major issue; many of the world's rivers are ecologically threatened as a result of the actions of humans who have polluted and damaged the supplies. The water is of poor quality and acts as a vector for water-borne diseases. Currently nearly 25% of the world's peoples lack access to safe water, so environmental sustainability protects water quality.

Economic sustainability involves guaranteeing security to water users from *all* groups at an affordable price. Interestingly, many schemes to manage rivers, such as mega-dams, actually dispossess people of their land, homes and livelihoods. Privatised schemes to bring clean, safe water to millions fail to deliver at affordable prices.

Economic sustainability is also achieved by minimising wastage and maximising efficiency of usage, for example in irrigation.

Sociocultural sustainability manages water supplies in such a way that it takes into account the views of *all* users, including the poor and disadvantaged peoples, and leads to equitable distribution within and between countries.

2.2 What physical factors determine the supply of water?

Figure 17 shows the hydrological system. It is the interactions between precipitation, evaporation, transpiration, surface runoff and groundwater transfers that determine the amount and availability of water supply.

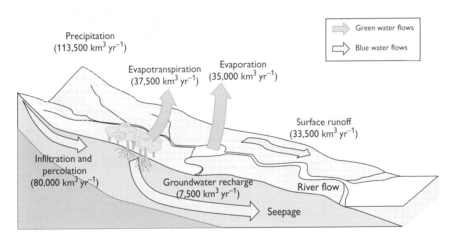

Figure 17 The hydrological system (global values)

A further complicating factor is that areas of high precipitation and plentiful water supply often do not coincide with the areas of greatest need.

There are a number of physical factors that influence the supply of water.

- At a macro scale, climate determines the global distribution of water supply by means of annual and seasonal distribution of precipitation (this includes rainfall and snowfall). Precipitation varies globally not only in terms of amount but also in terms of its seasonal distribution, availability and reliability — these last two characteristics are exacerbated by short-term climate change. The overall latitudinal distribution of precipitation belts is driven by the global atmospheric circulation. Areas of convergence and therefore uplift lead to precipitation in the equatorial zone and between about 45° and 60°, which are therefore areas of global surplus. Areas of divergence, for example between 20° and 40°, are semi-arid and arid areas with a global water deficit. The rates of evaporation and evapotranspiration are closely linked to temperatures, which can lead to periods of drought. Topography and distance from the sea have significant regional impact. High relief promotes increased precipitation and rapid runoff, but may provide greater opportunities for storage, especially where it is combined with impermeable geology.

- River systems. The world's major rivers store large quantities of water and transfer it across continents. The Amazon, for example, has an average annual discharge of 175,000 m^3 sec^{-1} from its catchment area of 6,915,000 km^2, shared by Brazil and six other South American countries. Recent droughts in the Amazon basin have therefore had a huge impact on water supply. Rivers that rise in the Himalayas

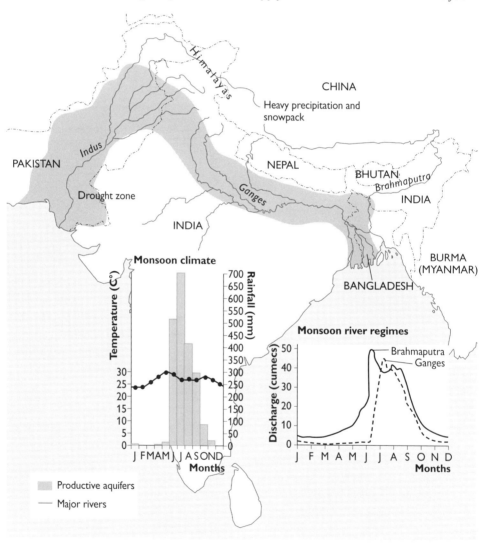

Figure 18 Factors influencing water supply in India

and the Andes are fed by melting ice from glaciers and snow. Global warming, which is melting glaciers and diminishing the snowfields, will ultimately impact on water security in countries such as Bolivia and Nepal.

- Geology controls the distribution of aquifers (water-bearing rocks) that provide groundwater storage. Permeable chalk and porous sandstones can store vast quantities of water underground, which is not subject to evaporation loss. The water comes from springs and wells and gives an even supply throughout the year, despite uneven distribution and variability of rainfall, provided they are not overused by rising demand at a faster rate than they can be replenished by natural recharge.

Figure 18 shows how these physical factors affect the water supply of India.

Water availability

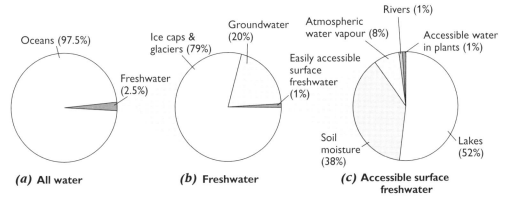

Figure 19 Water availability

Figure 19(a) summarises that although the volume of water in the world never changes, only 2.5% of it is freshwater. It is this percentage that is finite. Moreover, Figure 19(b) shows that two-thirds of this freshwater is currently unavailable for human use as it is locked up in ice caps and glaciers. Figure 19(c) shows how the easily accessible surface freshwater is held.

Human actions can have both negative and positive impacts on these stores, as shown by a case study of the Aral Sea (the world's fourth largest inland sea). By 2007 the sea had declined to just 10% of its original size — it had become an environmental catastrophe as a result of the damaging actions of humans. Only now is the Kazakhstan government beginning a scheme to restore the northern part of the Aral Sea. Restoring damaged wetlands is an ecological and sustainable way of compensating for the damaging actions of humans.

2.3 How do human activities influence water supply and demand?

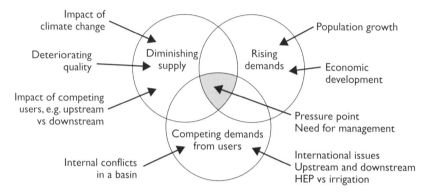

Figure 20 Water pressure points

Figure 20 shows how pressure or pinch points can result from the influence of human activities, which can lead to both diminishing supply and rising demands. Humans can impact on both the **quantity** and **quality** of the water supply.

Humans can remove water from rivers and groundwater sources, whether for drinking or other domestic uses, for irrigation or for industrial purposes (the three main sectors).

By 2025 total projected water withdrawals are predicted to reach over 5,000 km³ per year, of which agricultural use will be nearly two-thirds. Population growth and rising living standards, urbanisation, migration, economic development and industrialisation will have increased water demand to unsustainable levels in the pressure point areas and some new areas. **Over-abstraction** of supplies will occur if supplies cannot be replenished in time and reserves will be lost because rainfall can never fully recharge the underground stores.

Human actions can pollute both surface water and groundwater supplies, diminishing the quality of both sources.
- Untreated sewage disposal in developing countries causes water-borne diseases such as typhoid, cholera and hepatitis. As many people are forced to use unsafe water, it is estimated by WHO that, by 2020, 135 million people worldwide could die from water-borne diseases.
- Chemical fertilisers used increasingly by farmers (part of the Green Revolution) contaminate groundwater as well as rivers, and eutrophicate lakes and rivers, which leads to hypoxia and the formation of dead zones in seas.
- Industrial waste is dumped into rivers and oceans. Heavy metals and chemical waste (PCBs) are particularly toxic. (The Ganges is a useful example to study the full range of pollution, as are many rivers in China.)

- Many river management schemes, such as the construction of dams, have an impact on sediment movement that can impact on river ecology.

Figure 21 shows a model that summarises some of the human impacts on water supply and quality.

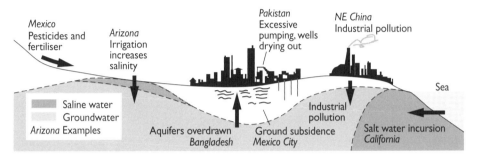

Figure 21 Human impacts on water supply and quality

Dwindling supplies and increasing demands can lead to pressure points, both within countries where there are conflicts between users (for example, the Colorado River) or between countries where the river-basin water is shared between countries (for example, the Tigris–Euphrates basin). Problems usually occur if upstream nations extract too much or return contaminated water to the system.

Table 7 identifies examples of the main surface water pressure points.

Table 7 Pressure points

Location	Reason for pressure point
Tigris–Euphrates basin	Concerns from Iraq and Syria that Turkey's South-Eastern Anatolian (GAP) Project will divert much of the water via a series of irrigation dams. Syria itself has developed dams too, which initially in the 1990s led to conflict with Iraq.
River Jordan	Use of the River Jordan, largely by Israel but also by Syria, Lebanon and Jordan, has reduced the flow of the river to a mere trickle. This affects supplies to Palestine's West Bank.
Ganges–Brahmaputra basin	India has built dams such as the Faraka, which has reduced the flow of the river into Bangladesh.
Syr Darya and Amu Darya Central Asia	Turkmenistan, Uzbekistan and Kazakhstan need more summer water for irrigation, but water has been diverted by schemes in Tajikistan/Kyrgyzstan.
Colorado basin	States in the USA dispute their allocation of water from the Colorado, which is so great that the quantity and quality of water reaching Mexico does not reach the standard agreed.
Nile basin	Although agreements exist, schemes developed in Ethiopia and Sudan may threaten supplies to Egypt.

Groundwater conflicts occur, especially in the middle east, Saudi Arabia and North Africa, where the water has been so over-abstracted that it cannot be replenished. In southwest USA, Spain and the Upper Indus, water is currently abstracted at a faster rate than it can be replenished. As many of the underground aquifers straddle international boundaries, there are complex issues to resolve shared groundwater usage.

- Supplies are underground so it is difficult to understand the problem as it takes years for any effects to show.
- Negotiating an equitable and reasonable share for each nation is difficult as nobody knows who owns what or how extensive the aquifers are as they have not been fully mapped.
- More powerful nations abstract groundwater more efficiently because they develop deeper wells and more technologically advanced pumps.
- United Nations' legislation to sort out water-sharing of aquifers is only just being written in 2010; this contrasts with the well-defined Helsinki Rules for surface-water use.

2.4 How can water supply and demand be managed sustainably?

Management of supplies

Increasing storage capacity and supplies

Approaches to increasing storage capacity range from high-tech fixes that involve building enormous dams to create huge new reservoirs to the use of traditional techniques which use intermediate technology to catch rainwater and store it in pumpkin tanks to provide water for local communities. These latter small-scale, bottom-up schemes tick many of the key features of sustainability, but the large-scale World Bank or government-financed schemes such as the Three Gorges dam, known as mega-dams, are far more controversial as sustainable sources of water supply.

Worldwide there are over 50,000 large dams, half of which are found in China. Nearly 60% of the world's major rivers have been dammed, in many cases fragmenting them by blocking their natural flow and turning them into a series of lakes. On some rivers, the volume of water that can be stored exceeds the annual flow at times, as in the case of the Colorado, leaving little water to reach the sea. Supporters of dams stress their multi-purpose benefits such as irrigation, flood control and HEP as well as water supply, whereas opponents refer to debt burden/costs, displacement and impoverishment of people and environmental destruction of important ecosystems and fisheries, all of which question their sustainability credentials.

In North America and Europe most technologically attractive sites have already been developed, but globally there are many dam projects under construction in

Turkey, Iraq, and Iran in the middle east, as well as Greece, Romania and Spain, and the Parana basin in South America. China and India, the two emergent superpowers, are also building many dams as a means of facilitating their economic development.

The World Commission on Dams concluded in 2004 that any dam project must achieve 'significant sustainable improvement in human welfare', i.e. it must be economically viable, socially equitable and environmentally beneficial — all elements of the sustainable stool if it were to be seen as the best solution. The Commission found that, in general, large dams did not achieve their water supply and irrigation targets, and that frequently environmental and economic costs were high. The Omo dams in Ethiopia, Three Gorges dam and the Euphrates–Tigris dams are all controversial examples (see **www.dams.org**). Many experts argue that, although there may be a need for further mega-dams in African countries such as DR Congo (often Chinese-built), the construction of small dams for single villages may be more sustainable. Inevitably, these projects lack prestige but locally these can be of enormous significance and value to communities.

New large reservoirs are also planned for areas such as southern England, which is increasingly an area of potential water stress. With large reservoirs there are issues of sustainability, particularly environmental, and many reservoirs are subject to huge losses from evaporation.

Water transfer schemes

Water transfers involve the diversion of water from one drainage basin to another (inter-basin transfer), either by diverting the river itself or by constructing a large canal to carry available water from an area of surplus to an area of deficit.

In the UK, with its generally wet northwest and much drier southeast — a factor potentially exacerbated by projected climate change — a water grid has long been planned but ruled out because of the costs of infrastructure, the energy-intensive need for pumping and, more recently, the difficulty of achieving cooperation between privatised water companies. With population expected to rise in southeast England by 1 million by 2025, in an already water-stressed area transfers may become essential.

Nevertheless there are numerous interregional pipeline transfers — from Welsh reservoirs to Liverpool and Birmingham, from the Lake District to Manchester and from Kielder Water to the River Tees and subsequently to the Yorkshire Ouse.

The controversy lies in large-scale, high-tech transfer schemes. The engineering itself and the actual water transfers have been successful, as Figure 22 shows, but there are many environmental and social disadvantages. Additionally, continuous use of transferred water may lead to long-term changes to local and regional hydrological conditions, perhaps increasing flood risk, damaging fishing stocks, spreading diseases and pollution and acting as a pathway for introducing alien species into new river environments.

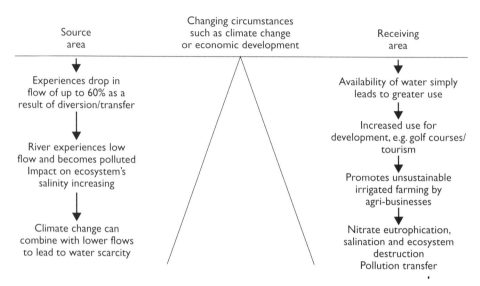

Figure 22 Water transfer issues

There are many large-scale water schemes in operation. Table 8 details some of these schemes and some of the proposed ones. These increasingly massive engineering schemes may be a techno-fix for water redistribution, but they have potentially huge environmental costs that challenge their claim to be sustainable. They may also be only a medium-term solution, so doubts exist over futurity. Equally, economic costs are so high that in countries such as Spain and Israel desalination may be a more viable alternative. They can also lead to conflicts about equitable use of water, both within and between nations.

Table 8 Examples of water transfer schemes: existing and proposed

Existing schemes	Details
Tagus–Murcia transfer in Spain	Takes water from Tagus River by canal to the drought-stricken area of Almeria/Murcia/Alicante to provide water for 700,000 new villas, the tourist industry and irrigated cash-cropping areas.
Snowy Mountains scheme in southeast Australia	Water is transferred from the storage lake of Eucumbene westwards by the Snowy tunnel to the head waters of the Murray River to irrigate farms and provide water to an increasingly drought-stricken area.
Melamchi Project in Nepal	Water is diverted from the Melamchi River via a 26 km tunnel to water-stressed areas in the Kathmandu basin. In return, the residents of Melamchi are provided with improved health and education services.
South to North Transfer Project in China	This project began in 2003 and will take 50 years to complete, costing up to $100 billion. It will transfer 44.8 billion m³ of water per year from the relatively water-secure south to the drought-stricken north via 1,300 km of canals linking the Yangtze to the Yellow Huai and Hai rivers.

Proposed schemes	Details
Ebro scheme in Spain	Following on from the Tagus scheme, 828 km of canals will be built to divert the waters of the Ebro to southern Spain.
Israel — transfer from any neighbours who would agree	Israel has a huge water deficit and plans for several schemes such as transferring waters from the Red Sea to top up the Dead Sea. Another scheme planned to tap surplus waters from the Mangarat River in Turkey, possibly by water tanker or under sea pipeline (subject to terrorism), but this has since been shelved.
Projected water transfer systems in Russia	Russia plans a whole series of schemes diverting rivers such as the Ob to the drought-stricken area of Aral Sea. The diversions could have major implications for the Arctic Ocean, as it would affect salinity.
Projected water transfer systems in India	India plans to develop a national water network to ensure better distribution of supplies to water-deficit areas such as the Deccan plateau.
Projected transfer projects in North America	Canada is a water-surplus country. NAWAPA is a scheme to take water from Alaska and northwest Canada to southern California and Mexico. A further scheme (Grand Canal) could take water from the Hudson Bay to the Great Lakes.

Groundwater sources and replenishment

The pressures of rapid population growth and economic development have combined with increasing uncertainty of water supply caused by short-term climate change to make the restoration of lost supplies from both rivers and groundwater a viable option.

In north London an artificial recharge scheme operates whereby treated water from sewage works is pumped into the chalk aquifers below London (the space is available because of the lowering of the water table from previous over-abstraction). In declared drought conditions (for example, summer 2005) the store can be used to supply boreholes in Enfield, Haringey, Walthamstow and Tottenham.

On Long Island, New York, where all the water supply comes from underground water reserves in sand and gravel aquifers, over-abstraction has lowered the water table and contaminated the groundwater to such an extent that this places limits on the area's growth. Legislation has now been passed so that any new developments must build recharge basins, proportional to their size, to collect runoff via sumps and storm water drains, as rain runs off the impermeable urbanised surfaces. This is clearly an example of sustainable development and a solution that is a way ahead for urbanised areas experiencing water supply vulnerability. Pumping water into natural aquifers for seasonal storage has a great future. In developing nations experiencing seasonal rainfall, this is the best solution as it is relatively cheap and not subject to evaporation losses.

Desalination

There has been a global boom in desalination, which draws on supplies from the oceans as opposed to the 1% of freshwater available for use on Earth. Therefore, it is a sustainable process as it conserves supplies for future generations. Although people have been desalinating water for centuries, recent breakthroughs in technology (for example, the development of the reverse osmosis process) have made desalination far more cost effective (given that freshwater exploitation costs are rising), less energy intensive and easier to implement on a large scale. However, it is still a costly option and does have a major ecological impact on marine life.

TNCs such as Veolia (France), Impergilo (Italy), Doosan (South Korea) and GE (USA) are building desalination plants around the world. The top six nations by desalination capacity are (greatest first) Saudi Arabia, USA, UAE, Spain, Kuwait and Japan.

With the future advent of carbon nanotube membranes requiring less pressure and therefore greater energy efficiency, relative costs of desalination may decrease still further, making this option — in effect the ultimate techno-fix — far more viable than massive hard-engineering water transfers. However, there are major concerns about its environmental impact because the left-over water returned from the desalination process has twice the salt concentration of sea water. Dumping it back near the shoreline will have adverse consequences on coral reefs and their food webs.

Exam hints

- Use books, articles and websites to research detailed examples and mini case studies that you can use to inform your arguments.
- Prepare a grid with the four facets of the sustainability quadrant — futurity, environment, public participation and equity — and devise a ranking or scoring system so that you can evaluate the degree of sustainability of the above four options for managing water supply.

Management of demands

With rising demands and dwindling supplies, **water conservation** is a key strategy. It is a sustainable strategy environmentally, and by most measures socioeconomically, as the aim is to conserve water supplies for the future. Each of the demand sectors can manage their use of water more sustainably.

Agricultural water demand

In **agriculture** the maxim has to be 'more crop per drop' where cash crops are grown. Sprinkler and surface flood irrigation systems are steadily being replaced by modern automated spray technology and more advanced drip irrigation that uses less water. Israel is a major pioneer of smart irrigation. There are also great savings to be made in repairing leaks in irrigation systems. As agriculture globally is by far the greatest user of water (just under 70%), it is here that conservation strategies are vital. As the world's population grows and incomes rise, it is inevitable that farmers will, if they use today's methods, need a great deal more water to feed the world's peoples.

For many farmers operating in areas of water scarcity such as northern China or western USA, there is a pressing need to make water go further. **Recycling** of city waste water for agricultural use is a feasible, relatively low-cost option as this **grey**

water does not need to be of drinking water quality. This recycling happens regularly on the North China plains.

Empowering farming communities to make their own decisions concerning water use has also been successful. There are numerous intermediate technology solutions to water conservation, such as the 'magic stones' system practised widely across the semi-arid Sahel or the development of devices to store and recycle rain in areas reliant on rains (for rain-fed agriculture). Experiments in Uzbekistan put farmers in control of the irrigation network and allowed them to decide how much water they needed, as opposed to giving them a fixed allocation. This cut consumption by 30%.

Specialised non-governmental organisations (NGOs) such as Farm Africa and WaterAid have helped farming communities develop a whole range of strategies to enable them to combat climate change induced water scarcity. Many schemes such as those in Ethiopia aim to provide better water storage of surplus water in the rainy season. They have also given farmers training in minimising tilling so that water is conserved in a layer of mulch on the field's surface, which absorbs the rainwater and limits evaporation. Agriculture advisers give guidance on types of crops that will generate good profits yet use less water, for example substituting dry crops such as olives for thirsty citrus fruits. A current controversy is the thirstiness of crops used to produce supposedly sustainable biofuels such as ethanol and biodiesel.

High technology also has a role to play. Second generation GM crops are being bred that are not only tolerant of diseases but also of drought and salty conditions — these include strains of maize, millet and wheat, which are vital food crops.

The Murray Darling basin, often called the Food Bowl of Australia, is in crisis as a result of historically high use for agriculture, climate change and recent prolonged droughts. A basin-wide sustainable management plan aims to reduce surface and groundwater use by 10% by limiting the amount that can be taken. Buying land from farmers to allow natural runoff to the river and environmental watering plans will restore the wetlands as a water store.

Agronomists are also beginning to devise tools to help monitor the efficiency of water use. Some have designed algorithms that use satellite data on surface temperatures to calculate the rate at which plants are absorbing and transpiring water — this means development agencies can concentrate their efforts for improvement on the most thirsty crops.

Systems have been devised to grow crops using little water — **hydroponics** involves growing crops in huge greenhouses that are carbon dioxide and temperature-controlled. There is no soil and crops are grown in shallow trays where they are drip-fed nutrients and water. The only issue is that it may be a sustainable system water-wise suitable for supplying food to arid lands, but it is very energy intensive.

Water demand in the industrial sector
Industry is the next largest user of water, using some 20% of total water withdrawals. For businesses, water is not discretionary as 'without water, industry and the global

economy falter'. Water is an essential ingredient of many food and beverage products such as beer and soft drinks. It is also used in a huge range of other industries such as making silicon chips or for cooling thermal power stations.

Rapid industrialisation, particularly in developing countries, has contaminated both rivers and aquifers and for many industries it is not so much the quantity of water but the quality of the water that is important.

Many large TNCs have reduced their consumption of water; for example, Coca-Cola bottling plants around the world are committed to clean all their waste water by 2010 and then to **recycle** some of it for use as grey water in their plants for cleaning bottles and machinery. Coca-Cola is currently the largest beverage company in the world and has been the subject of adverse publicity over its intensive use of water (283 billion litres worldwide, 4 litres of water needed for 1 litre of product), especially in Kerala in southern India. Another strategy used by Coca-Cola as part of its environmental mission is to provide training and technology to help villagers conserve rainwater and irrigate crops more efficiently.

Many companies have improved their **recycling** of water as a response to legislation prohibiting use of groundwater or to rising costs in the price of water. In Beijing in water-stressed northern China, there are zero liquid discharge rules that ban companies from dumping waste water into the environment, which forces companies to recycle all their waste water by treating it by purification for reuse as grey water.

One issue that is becoming a hot topic is the concept of **virtual water**. The industrialised lifestyles of developed countries benefit from imported as well as locally produced goods, which involves the consumption of 'virtual' water embedded in the food and manufactured items. Figure 23 illustrates four contrasting **water footprints** and adds together water consumed directly and indirectly in imported products. This concept emphasises how unsustainable the use of water is in most developed countries, as well as in many OPEC and South American countries. Some nations see outsourcing food production and therefore the use of water as a future option but it is certainly not sustainable — it merely transfers the issue.

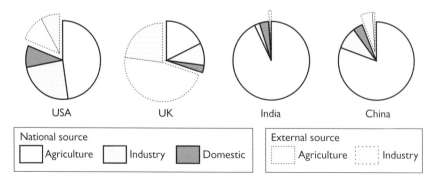

Figure 23 Proportion of embedded water by source, 2001

Domestic water use

The third major sector for water use is described as **domestic**, which currently uses around 8% of total withdrawals. Globally there is a huge divide between developed nations, which use up to five times the cubic metres per capita compared with many developing nations.

Water conservation is a key strategy — more of an attitudinal fix than a techno-fix. Domestic water conservation includes reducing consumption by the installation of smart meters, which can monitor use and make higher charges in stress periods such as dry summers. Rain harvesting using a system of rain butts is a further conservation measure in the garden. Strategies such as sharing a bath, putting a brick in the toilet cistern or using an eco-kettle can cut down on consumption, often inspired by the threat of rising costs of metered water. In times of drought, water conservation can be enforced by hosepipe and sprinkler bans. Recycled water can be encouraged for flushing the toilet or garden use such as watering plants with left-over washing-up water.

Filtration technology now means that **physically** there is little dirty water that cannot be purified and recycled. Faced with the loss of cheap imports of water from Malaysia, Singapore has followed a path to water self-sufficiency — artificial rain catchments combined with treating sewage water. Water cleaned by a combination of dual membrane technology (micro filtration and reverse osmosis) and ultraviolet disinfection produces water that exceeds WHO quality thresholds. It is marketed as Newater and is now a key source of supply for the densely populated island state (in 2010 some 30% of Singapore's drinking water was Newater). However, a psychological barrier has to be overcome on drinking water from toilets!

Technology can be useful in a number of ways, such as water companies carrying out projects to cut down on leakage from broken pipes and burst water mains or treating and reusing industrial and waste water at their waterworks. Another development concerns the construction of climate-proofed gardens filled with drought-resistant species that can survive periods of water stress.

The main strategies for reducing demands for water — i.e. recycling, grey water use and reducing consumption — can clearly make a major contribution to the sustainable management of water. The question is, are all these strategies completely sustainable?

2.5 Can sustainable water supplies be maintained in the future?

Figure 24 shows that total water withdrawals are predicted to reach nearly 5,000 km³ per year by 2025, with a considerable impact on ecosystems and people's health and wellbeing. There will be a disproportionate knock-on effect on the lives of the world's poorest peoples, who will suffer from water and food insecurity. In 2002 the

International Food Policy and Research Institute (IFPRI) used computer models to examine the implications of three alternative futures for global water supply and demand (Table 9).

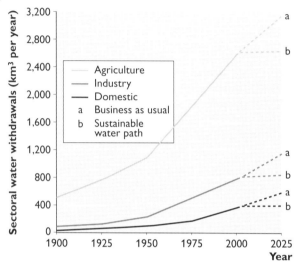

Figure 24 Rising trends in water use

Table 9 Alternative scenarios for water by 2025

Scenario	Water changes by 2025	Wider impacts
Business as usual	• Water scarcity will reduce food production. • Consumption of water will rise by over 50%. • Household water use will increase by 70% (mostly in developing countries). • Industrial water demand will increase in developing countries.	• Developing countries will become reliant on food imports and experience increased hunger and malnutrition. • In sub-Saharan Africa, grain imports will more than triple. • In parts of western USA, China, India, Egypt and North Africa, users will pump water faster than aquifers can recharge.
Water crisis	• Global water consumption would increase further, mostly going to irrigation. • Worldwide demand for domestic water would fall. • Demand for industrial water would increase by 33% over business-as-usual levels, yet industrial output would remain the same.	• Food production would decline and food prices, especially of cereals, would increase rapidly. • In developing countries malnutrition and food insecurity would increase. • Dam building would decline because of fewer potential sites and key aquifers in China, India and North Africa would fail. • Conflict over water between and within countries would increase.

Sustainable water	• Global water consumption and industrial water use would have to fall considerably. • Environmental flows could be increased dramatically compared with other scenarios. • Global rain-fed crop yields could increase due to improvements in water harvesting and use of sustainable farming techniques. • Agricultural and household water prices might have to double in developed countries and triple in the developing world.	• Food production could increase slightly and shifts occur in where it is grown. • Prices could fall slowly. • Governments, international donors and farmers would need to increase investment in crop research, technology and reforms in water management. • Excessive pumping is unsustainable. • Governments could delegate farm management to community groups.

The business-as-usual scenario is unsustainable in the long term. The most worrying scenario is that of water crisis, which shows how mismanagement of water resources or climate change could threaten supplies and lead to wider environmental problems and conflicts. The sustainable water scenario is essential for a water-secure future, but it involves some radical solutions that may not be acceptable to all the players involved in water use.

A number of questions remain:

- Social players see access to clean, safe water as a right. Inevitably the UN's MDGs will cost a huge amount of money — around $250 billion to halve the proportion of population without access to improved water supply and sanitation. Will the developed world fund this amount to supply pipes, sewage systems and water treatment plants? Governments, especially in low-income countries, cannot afford these huge sums, so will the investment need to be made by transnational privatised profit-led organisations that have the capability and technology to do the work but will need to be offered favourable terms to invest? In Bolivia and Tanzania rising prices for consumers take the water provided well above the price communities can afford.
- Many governments and businesses favour high-tech, top-down hard engineering solutions to improving the world's water supplies, but will bottom-up, non-NGO community-based projects favoured by social and environmental players be ultimately more successful and more sustainable?
- Can global frameworks for equitable water use and management of the global pot be established regionally (as for climate change)? Which national and local governments and communities can work out guidelines for sustainable water use?
- Can cooperation between nations be established to share and manage the equitable use of the dwindling supplies, therefore avoiding the prediction of 'water wars'?

- Can the water gap be bridged — with more equalisation of water footprints? An obvious manifestation of profit-legislated and unsustainable use in developed countries is the issue of bottled water.

Is integrated water resources management (IWRM) the way ahead?

Figure 25 shows the main features of IWRM designed to:
- protect water quality
- ensure economical usage
- distribute water equitably

Figure 25 The main features of IWRM

IWRM encompasses many aspects of sustainable water management of supplies and demands. The physical resource is the starting point. Satellite images and water accounting are used to determine how much water there is, how productively it is currently used and how this could be improved. Of paramount importance is the way the water is managed within a basin community and the impact this makes on the surrounding environment.

> The village of Thunthi Kankasiya in Gujarat, India is an example of locally successful IWRM where the bottom-up approach is designed to overcome water poverty. In the village appropriate technology schemes now provide a year-round water supply, enabling triple cropping and the quadrupling of the production per hectare. This raised annual household incomes five-fold, well above the Indian average.

Source: *The Atlas of Water*, Earthscan, 2009

Figure 26 shows the process of IWRM.

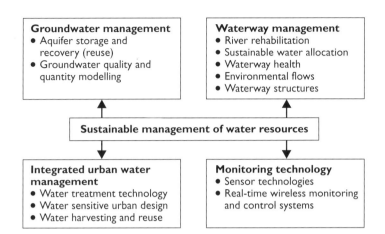

Figure 26 Integrated sustainable water resource management

Theme 3: Sustainable energy

3.1 The issues

Energy is fundamental to human existence as it is needed for transport, making a living and to achieve a satisfactory quality of life. One of the greatest challenges facing humanity in the twenty-first century is ensuring that everyone has access to safe, clean and affordable energy supplies. Although fossil fuels provided the power for the first Industrial Revolution (nineteenth-century coal-based) and established the wealthy first world, it was oil and associated gas that continued the wealth generation in the twentieth century. There is, however, a growing realisation that fossil fuels are ultimately finite and an increasing concern that supplies could run out — at current levels of consumption coal has a resource/production (R/P) ratio of 200 years, gas 60 years and oil only 40 years — and that future supplies of energy need to be sustainable. Moreover, as a result of natural processes fossil fuels (especially oil) are geographically concentrated (the middle east and North Africa have 70% of the world's oil and gas supplies), threatening energy security for many parts of the world for physical, environmental, economic and geo-political reasons.

The massive use of coal, oil and gas has fuelled enormous increases in material prosperity — at least for most developed countries and increasing numbers of NICs, which now include the emergent superpowers of India and China. However, this consumption has had numerous, largely environmental, adverse consequences including air and water pollution, coal-mining accidents (in present-day China), fires, explosions and tanker accidents. Most scientists now consider that present-day global climate change is anthropogenically caused and the warming is likely to be linked to the increasing atmospheric carbon dioxide and other greenhouse gas concentrations caused by fossil fuel combustion.

The sustainability quadrant (see page 12) helps us to consider what are the main features of energy sustainability. Figure 27 summarises these features.

Futurity	Environment
Energy use in which sources overall are not depleted by continued use.	Energy use that does not entail the emission of pollutants or other hazards to the environment on a substantial scale.
Public participation	**Equity and social justice**
Energy use that involves the community in decisions about the appropriate choice and use of energy mix.	Energy use that does not involve the perpetuation of health hazards or social injustice (for example, dispossession of land of native people) and does not prevent poor people and poor communities gaining access to adequate supplies at affordable costs.

Figure 27 The energy sustainability quadrant

In summary, sustainable energy policy should aim to:
- implement greatly improved technology for harnessing fossil fuels and nuclear power, to ensure a much reduced environmental impact
- develop and deploy renewables on a much wider scale using technology to overcome issues of feasibility and high costs and make their benefits accessible to all countries
- improve efficiency of energy distribution and use, including developing effective conservation strategies

3.2 What problems are associated with the supply of energy?

Types of energy supply

Energy resources can be classified as non-renewable, renewable and recyclable.
- Non-renewable energy sources (also known as finite, stock or capital resources) have been built up over millions of years, but they cannot be used without depleting the stock because their rate of formation is so slow. They include fossil fuels as well as nuclear energy using fission processes. Nuclear power is an alternative energy source to fossil fuels; it does not emit carbon dioxide but there are other environmental concerns about radioactive uranium fuel and the long-term problem of disposing of nuclear waste.
- Renewable energy sources (also known as flow or income resources) yield a continuous flow that can be consumed in any given period of time without

endangering future consumption as long as the amount used does not exceed net renewal during the same period. Renewable resources can be subdivided into two groups.

- Critical resources are recyclable and therefore sustainable. They include energy resources from forests, plants and animal waste. Nuclear power can also be considered recyclable when produced by fusion from recycled waste fuel. Carbon dioxide is emitted from biomass/biofuels but reabsorbed when they are 're-grown', making them potentially carbon neutral.
- Non-critical resources are everlasting resources such as tides, waves, running water (for HEP), wind and solar power. These resources produce no carbon dioxide and do not contribute directly to atmosphere pollution.

Table 10 summarises the main characteristics of commercial **primary energy sources**.

Table 10 Main characteristics of commercial primary energy sources

Primary energy type	Classification	% of global energy supply, 2008	Key concerns/issues
Crude oil (petroleum) A naturally occurring mineral oil consisting of many types of hydrocarbons. Crude oil may include small amounts of non-hydrocarbons. Also includes tar sands and oil shale.	Non-renewable	37%	Concerns that global supplies may have reached their peak, security of supply, geopolitical tensions and lack of alternatives, especially for transport. Releases carbon dioxide when burnt.
Coal A combustible, sedimentary rock formed of converted residual plant matter and solidified below overlying rock strata. There are several types of raw coal: hard/ bituminous coal, brown coal (lignite) and peat.	Non-renewable	25%	Use releases large amounts of carbon dioxide and other pollutants, contributing to climate change and atmospheric pollution. Carbon-capture technology for removing carbon dioxide from atmosphere unproven and complex.
Natural gas A methane-rich gas found underground. It also contains water vapour, sulphur compounds and other non-hydrogen gases such as carbon dioxide, nitrogen and helium.	Non-renewable	23%	Costs and security of supply, especially for countries that are largely importers, e.g. UK. Releases carbon dioxide on use.

Primary energy type	Classification	% of global energy supply, 2008	Key concerns/issues
Nuclear fission The division of a heavy nucleus into two parts, usually accompanied by the emission of neutrons (neutral particles inside the nucleus), gamma radiation (high-energy radiation) and energy release. This energy is converted into heat that raises steam to drive turbines and generate electricity.	Non-renewable (may be recyclable) with fuel reprocessing	6%	Possible health risks associated with power plants and accidents such as Chernobyl. Disposal of radioactive material raises safety issues and there are unknown long-term risks. Amount of raw material left globally is finite.
Biomass Organic, non-fossil material of biological origin. Although the different forms of energy from biomass are considered renewable, their rates of renewability differ. Wood is an example of a biomass energy source.	Recyclable	4%	Relatively low energy densities mean limited potential for large-scale electricity generation. Biomass acts as a carbon sink, so combustion releases stored carbon dioxide. Very efficient domestic wood chip burners now in use.
Hydrological energy (HEP) Energy harnessed from the movement of water through rivers, lakes and dams (owing to gravity). A 'head' of water is stored and then released to drive turbines and generate electricity. HEP systems can range in capacity from thousands of megawatts to small micro-hydro schemes.	Recyclable and renewable	3%	Large-scale systems are costly to build. Dam-building also has social, political and environmental impacts. Smaller micro-hydro plants may not be economically viable.
Ocean energy Energy harnessed by using either the physical characteristics of oceans (tidal movements, wave motion, thermal gradients, ocean currents) or their chemical characteristics (saline gradients).	Renewable	0.8%	Only certain locations, e.g. Alderney, are suitable for offshore tidal generation. Technology for large-scale generation is unproven. Ocean sources have low energy densities and large devices are needed to harness this energy.

Primary energy type	Classification	% of global energy supply, 2008	Key concerns/issues
Solar energy Energy directly harnessed from solar radiation, as distinct from wind, water and biomass energies indirectly driven by the sun. Solar radiation is absorbed by a collector and converted to heat energy or into electricity by photovoltaic (PV) cells.	Renewable	0.5%	Distribution and availability varies spatially and temporally. Photovoltaic technology still expensive compared with fossil fuels.
Wind energy Directly related to solar activity, which causes differences in atmospheric pressure and temperature (and in Earth's rotation and gravity). Modern wind turbines range from 600 kW to 5 MW of rated power.	Renewable	0.5%	Only certain locations have enough wind to be viable. Wind energy is variable, so it is difficult to manage power supply through a grid system without some back-up.
Geothermal energy Comes from rocks within the Earth and can be tapped in three ways: (1) as hot water or steam, (2) as hot dry rock energy, (3) by means of conduction. The first two are used to generate electricity whereas the third is used to heat water, buildings and greenhouses.	Renewable	0.2%	Geothermal heat in the outer 10 km of the Earth's crust is too diffuse to be exploitable worldwide. Availability is limited to a few locations such as central Southampton, Iceland and the Philippines.

Raw materials are used for power in their natural form such as coal, oil, uranium and wood (mainly non-commercial use). These primary resources can be converted into **secondary energy sources** such as electricity — hence the low percentage of nuclear (6%) when it contributes 20% of electricity globally. The main downside of electricity is that it cannot be stored, so demand and supply have to be in unison. This has implications when countries try to achieve the appropriate energy mix for supply.

Problems associated with energy supply

Energy costs

Oil and natural gas
It is likely that for the next few decades fossil fuels will continue to supply most of the world's energy needs (currently nearly 85% of primary energy demands), but

rising costs, especially of oil and gas, are a major concern. Oil prices have fluctuated considerably in the last decade, rising to a high of around $140 dollars a barrel in autumn 2008 before falling demand set in from the global recession and prices dropped to below $40 a barrel. Demand remains high for oil and gas as they are flexible and relatively clean compared with coal, and more energy dense. They can be transported efficiently by supertankers and pipeline. The big concern is that there is a possibility we may have reached **peak oil** (the point at which the maximum rate of global oil production is reached), with the prospect of declining production and rising prices (Figure 28). This depends on exploration and subsequent discoveries.

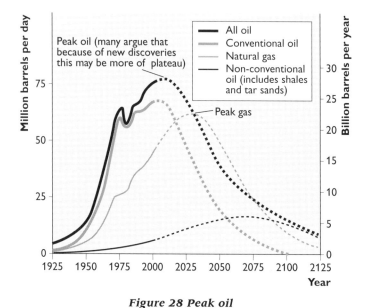

Figure 28 Peak oil

The scale and pace of exploration by TNCs and governments is closely tied to oil prices. With high prices, unconventional sources of oil such as tar sands and oil shales become viable or desirable, as do conventional sources in difficult deep-water locations (for example, offshore Brazil and the Gulf of Mexico) or in environmentally sensitive areas (for example, Alaska). Not surprisingly, the timing of reaching peak oil state is hotly disputed.

- In 2007 the German Energy Watch group claimed the peak was reached in 2006.
- In 2008 the UK Industry Task Force on Peak Oil and Energy Security suggested 2013 as the key date.
- In 2006 the IMF predicted oil production rising to 120 million barrels per day by 2030 with no peak yet reached.
- Yet other groups suggest it should be called plateau oil, with a steadily managed, OPEC-controlled production at around 80–85 million barrels per day.

The nearer the peak being reached causes the price of oil to rocket as so much of its production is concentrated in politically unstable areas. Other analysts suggest that it is a lack of oil refinery capacity, combined with stock market speculation, that is causing the price hikes.

Note that there is less concern about peak gas, which will possibly be reached between 2025 and 2030, especially with the advent of new supplies of shale gas.

Coal

On the other hand, coal is an abundant fuel, mined across the world with reserves that will last for at least 200 years at current use rates. However, there are a number of environmental and social issues related to the cost of coal. Highly mechanised open-cast mining, such as in the huge lignite pits in Germany and Kentucky, USA, may be the lowest economic cost, but the environmental costs are enormous. There are huge social costs associated with deep mining in countries such as China, where up to 3,500 miners are killed each year in accidents. Coal is less energy dense, more costly to transport even using new slurry technology, and dirtier — especially in terms of causing acid rain containing sulphur dioxide and nitrogen oxide.

Nuclear power

A large question mark hangs over the costs of nuclear power. The research and development costs are enormous, as are the construction costs of nuclear reactors in order to ensure health and safety, often taking some 10 years to get planning permission before they can be built. A single large nuclear reactor produces the power equivalent of 600 large wind turbines, so figures can be produced to show that in terms of running costs only nuclear power is cost competitive and can make a fundamental contribution to base-load electricity. However, of the 439 nuclear reactors operating globally in 2008, only 34 were constructed in the period 1998–2008, and nearly half are between 20 and 30 years old and therefore ready for decommissioning. Although nuclear power has the ability to bridge the energy supply gap, as an alternative to diminishing oil and gas supplies, 'dirty' coal and the large-scale development of renewables, unless a huge programme of nuclear-capacity building is launched soon, nuclear power may decline in importance, causing electricity shortages in some countries. There are many controversies surrounding nuclear power, including radioactive fuel, terrorism concerns, waste disposal and the technology linkage to nuclear weapons (a big issue with Iran).

Ocean energy and wind power

The cost of renewables is dropping steadily, but costs remain high as many are currently at a developmental phase (waves and tidal) or incapable of delivering large amounts of electricity to compete with large thermal or nuclear power stations. Renewables are 'clean' and environmentally friendly in terms of lack of carbon dioxide emissions, but they do have other environmental costs that lead to fierce outbursts of local Nimbyism. A case in point is wind farms: in the UK, 80% of people support the development of this form of energy, while at the same time protesting against their local construction. Unfortunately many of the best sites are on exposed coasts or hilltops — highly visible in some of the most beautiful places in the UK. This therefore pushes the building of larger, efficient turbines in bigger wind farms to offshore sites such as Horns Rev in the North Sea to overcome concerns about the intermittent nature of wind power.

HEP

HEP is another clean renewable source, but mega-dams are increasingly controversial. Cost-wise, dams may be justified by their multi-purpose use for irrigation, tourism and flood control, but with many recent sites there are negative environmental costs. The Omo dam in Ethiopia shows how controversial is Ethiopia's desire to be an HEP exporter to surrounding countries; not only will the dam threaten the livelihoods of many tribal people, it will also add to water problems in the neighbouring Turkana area of northern Kenya.

Solar power

Solar power is seen by many as a potentially cost-effective way of generating power. Both China and India are investing billions in developing technology that aims to lower the costs and raise the conversion efficiency of PV cells. The big question is, can it be successfully up-scaled? A recent development is outsourcing supply to the hugely sunny parts of the world such as the Sahara Desert and southern Spain. However, the problem is that, as with HEP, many of the sites are great distances from major centres of population and involve the development of high cost and ugly transmission lines.

The development costs of most alternative sources are so high and the technology so complex that large-scale developments are beyond developing countries needing more power for their emergent economies and costs.

> **Exam hint**
>
> Prepare a table in which you summarise all the major sources of primary energy. Assess their degree of environmental, economic and social sustainability as large-scale providers using Figure 27.

Climate change impact

Only with the development of carbon capture and storage (CCS), which involves capturing carbon dioxide released by burning coal and burying it deep underground, can clean electricity be produced from coal. There are, however, huge problems with the technology, as it is extremely expensive and only at the pilot stage of development — no one knows for sure whether it will work or even whether the carbon dioxide will stay trapped underground.

Figure 29 explains why nuclear power is favoured so strongly by environmentalists — because all forms of energy that depend on fossil fuels lead to carbon output, even the supposedly cleaner gas. However, coal power stations can be built quickly and cheaply (China commissions a new one each week) so this is a tempting supply solution. In South Africa, which is currently experiencing a supply problem, the World Bank is deciding whether to fund an enormous coal-fired power station as the most realistic solution to the shortages of power.

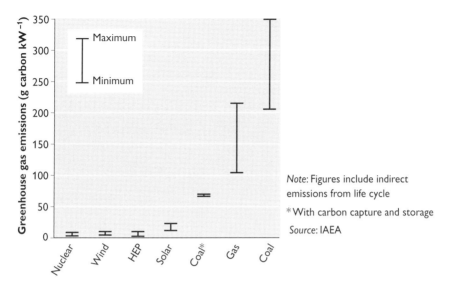

Figure 29 Greenhouse gas emissions of energy sources

As the complex negotiations at Bali (2007) and Copenhagen (2009) show, the development of a binding agreement to control levels of carbon dioxide emissions is fraught with controversy at a number of levels, including the fundamental one as to whether the recent rise in temperature (global warming) is definitely linked to a rise in carbon dioxide emissions. The talks also exposed huge tensions between developed nations, emerging superpowers such as China and India and developing nations as to how best to mitigate carbon dioxide emissions. The big issue is that all the clean technology comes with enormous costs and, as we saw when evaluating costs of supply, you may begin to solve the greenhouse gas emission problem but all alternatives have major environmental implications. Climate change environmentalists are so opposed to new coal-fired power stations in the UK and Germany that new builds are almost impossible to implement as the level of protest is massive.

Energy security

It is in every country's interest to be as energy secure as possible. The key to energy security success lies in a number of criteria:

- Making the greatest possible use of domestic sources of energy, sometimes at a higher economic cost or incurring major environmental costs (US and Arctic oil drilling and tar sands exploitation).
- Diversifying energy resources to minimise dependency on imported oil and gas and at the same time maximising the use of a range of alternatives (possibly nuclear) and cost-effective renewables. This requires substantial investment in technologies.
- Ensuring guarantees of imported energy by using reliable supplies with safe international pathways so that prices remain as stable as possible. Long-term

agreements are vital with 'friendly' suppliers and a variety of supply routes; routes can be disrupted by war, political issues, piracy (Somalia) or conflict, such as Russia's dispute over gas prices with Ukraine in 2008 and Belarus in 2009, which had knock-on effects because of pipeline disruption for most of continental Europe. The new Nabucco and Nord and South stream pipelines, are proposed alternative routes.

- Ensuring coordination and planning of future supplies — in the case of the UK, of the privatised energy companies. This is needed to ensure adequate storage of gas and the appropriate mix of sources to ensure both present and future needs by avoiding an energy supply gap. This gap has been caused by under-investment in new plant, as experienced in South Africa in 2008–09, and will be of concern in the UK around 2016 when coal power stations are shut and ageing nuclear plants decommissioned.

Figure 30 shows the energy security square, which identifies the key factors.

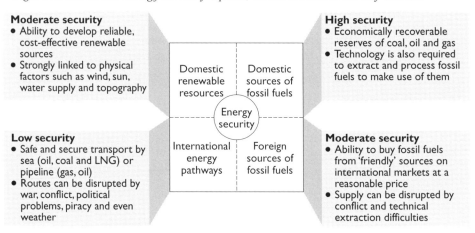

Moderate security
- Ability to develop reliable, cost-effective renewable sources
- Strongly linked to physical factors such as wind, sun, water supply and topography

High security
- Economically recoverable reserves of coal, oil and gas
- Technology is also required to extract and process fossil fuels to make use of them

Low security
- Safe and secure transport by sea (oil, coal and LNG) or pipeline (gas, oil)
- Routes can be disrupted by war, conflict, political problems, piracy and even weather

Moderate security
- Ability to buy fossil fuels from 'friendly' sources on international markets at a reasonable price
- Supply can be disrupted by conflict and technical extraction difficulties

Domestic renewable resources | Domestic sources of fossil fuels

Energy security

International energy pathways | Foreign sources of fossil fuels

Figure 30 The energy security square

Any potential or actual threats to disrupt oil supplies cause oil prices to rocket because of the geopolitical tensions associated with oil. These high prices can lead to political instability and economic stagnation as oil provides the bedrock for modern living.

The energy development gap

In developing countries, particularly in rural areas, 2.5 billion people rely on biomass, largely fuelwood but also dung, to meet their energy needs especially for heat and cooking. In countries in sub-Saharan Africa, fuel for cooking accounts for around 90% of household energy consumption because until recently (with the advent of micro-hydros and solar kits) most rural dwellers had no access to electricity. In areas of high density population, the use of this recyclable resource has become unsustainable. Not only are people facing scarcity of supply, but the removal of the protecting vegetation cover has led to widespread soil erosion. It has been described as the 'other people's energy crisis' (as opposed to oil in developed countries).

There are also human costs of fuelwood use; using wood-burning stoves in confined spaces is one of the greatest causes of ill health. Impaired lung capacity and cancer from smoke particles contribute to low life expectancy. Women and children bear the burden of fuelwood collection — a time-consuming and exhausting task as the length of journey to collect the diminishing supplies may be a round trip of 10 km per day. The women are prevented from food production tasks and children, especially girls, miss out on school. Fuel-efficient stoves such as the Jiko and more energy-intensive green charcoal pellets, as well as tree nursery and afforestation schemes, are helping to improve this situation.

Technology that requires a raft of educated people, research facilities and funding for research and development is bypassing many developing countries, so widening the development gap. Almost all large-scale energy developments would therefore need to be supported by multi-lateral aid (from the World Bank) or foreign direct investment (from TNCs). As with water supplies, the way ahead will include life-enhancing intermediate energy technologies that are appropriate to rural areas in developing countries. China and India are both helping to bridge the energy gap with massive investment, research and development, and production of renewable energy technologies.

3.3 How and why is the demand for energy changing?

Energy is used for a wide variety of purposes, from generating electricity for powering industry and homes to propelling transport. Figure 31 shows the demand patterns for all the main fuels.

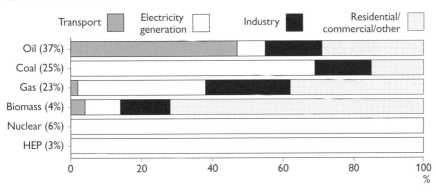

Figure 31 Global energy use, 2005

Note how some primary sources such as nuclear and HEP are used totally for secondary use, i.e. electricity generation, whereas nearly half of the world's oil is used for transport.

One feature of the demand is **rising consumption**, which is shown in Figure 32.

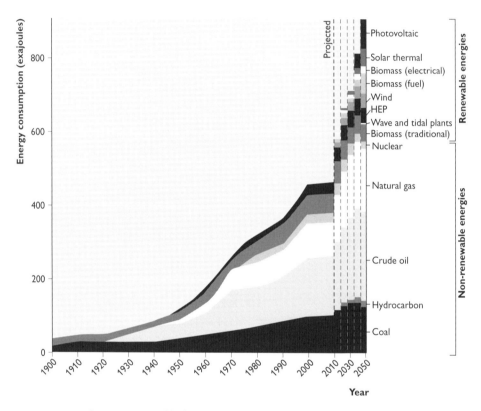

Figure 32 Worldwide energy consumption, 1900–2050

As Figure 32 shows, during the twentieth century energy demand increased tenfold, but by 2050 consumption is expected to double to around 900 exajoules (an exajoule is 10^{18} joules). This rising demand is linked to a number of factors:

- Economic growth depends on energy. This is true of the NICs, especially the large countries of Brazil, Russia, India and China (collectively known as BRIC) where energy for manufacturing is an important driver of growth. Economies such as China's grow at around 10% per year. China is known as the 'Workshop of the World' and this exponential growth requires huge quantities of energy.
- As countries develop economically, their people acquire more wealth and therefore more energy to use appliances and gadgets for cooking, heating, air conditioning and lighting. There is a gradual transition from poverty to the development of a society with a strong middle-class element who enjoy comfortable lifestyles (as in Brazil).

- Labour-saving devices, together with industrial legislation cutting work hours and increasing paid holidays, free up time for leisure and social activities. Car ownership becomes an aspiration and is rising rapidly in countries such as China, as people want to get to work in comfort, possibly commuting significant distances, and to travel to see friends and relatives and to enjoy hobbies and holidays. The 1.4 billion Chinese have discovered the freedom of car trips and internal tourism flourishes.
- Many modern cities are designed to be low density and sprawling and are built with the car in mind (see page 80).
- In an increasingly globalised world, with growing international tourism and trade, the transport of people and goods by air, sea and land has increased enormously — hence the concern about food miles (see pages 33–34) and carbon footprints.

There is therefore a strong positive correlation between GDP per capita and energy usage.

Figure 32 also shows how the energy mix is expected to change in the next 40 years. Although the graph predicts a post-2030 increase in renewable energies, especially solar, there is still an overwhelming dependence on non-renewables, especially fossil fuels. Note, however, a slight decrease in their consumption after 2040.

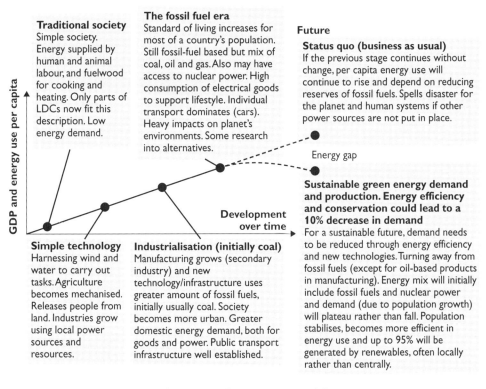

Figure 33 The energy transition

Estimates in energy consumption depend on whether a 'business as usual' or 'more sustainable' policy is adopted, whereby the growth in demand could be restrained by about 10%. To do this, governments and other decision makers would have to deliver cleaner and more efficient energy supplies together with a push towards energy conservation.

Figure 33 summarises the energy transition that countries undergo when they develop economically.

3.4 How can the demand for energy be managed sustainably?

Energy in a sustainable future involves three key strategies:
- Cleaning up fossil and nuclear fuels.
- Switching to renewables using appropriate technology.
- Using energy more efficiently by developing conservation strategies by which it should be possible to deliver environmental and socio-economic sustainability and conserve scarce resources for future generations.

Cleaning up

As concerns grow about the impacts of climate change and reaching the tipping point that will lead to the catastrophic and irreversible collapse of the world's environmental systems, there is huge pressure to develop **clean coal** technologies. Coal is the most ubiquitous fossil fuel with the greatest reserves (resource:production ratio of about 200 years).

Pollutant emissions of sulphur dioxide and nitrogen oxide and particulates from coal combustion can be substantially reduced using techniques such as fluidised bed combustion and flue gas desulphurisation (using scrubbers). However, carbon dioxide emissions are the most difficult problem to solve. **Carbon capture** and **carbon sequestration** of atmospheric carbon, either in forests or below the Earth's surface, are the main potential solutions.
- Carbon sequestration can be achieved by growing additional forests. These may have added potential if they can be used for biofuels or are leguminous trees, which would add nitrogen to the soil.
- Carbon sequestration could take place by injecting carbon dioxide into the oceans or by enhancing natural ocean uptake of carbon dioxide but this is only under investigation.
- Carbon capture and storage is one possible solution. This is at the pilot phase in the UK, with projects in southern Germany at the operation stage.
- Carbon dioxide can also be sequestered in deep beds of coal that are too inaccessible to mine. These coal beds usually contain significant quantities of methane, which could be piped to the surface and used as a potential fuel (coal gas). Coal or oil can be used to produce syngas for use in gas-fired power stations.

Accepting that coal is the dirtiest fossil fuel (Figure 29), a solution frequently used is to change power generation and fuel use to **natural gas** — the cleanest (lowest carbon content) of the three fossil fuels. This is known in the UK as 'the dash for gas' as all recently built power stations have been gas fired using combined cycle gas turbine plants. Domestically, natural gas can now be burned in condensing boilers that are extremely efficient — over 90%. Biomass has been burned in conventional power stations at Drax since 2003. It uses willow and straw from the largest source of renewable energy in the UK.

As far as cleaning up **oil** is concerned, one priority is to use sulphur-free oils. There are designs to reduce pollutant emissions from fossil-fuel combustion, including particulate filters on diesel engines, clean lean-burn engines and more sophisticated engine management systems. In the future, a fossil-fuel based hydrogen economy may be the way ahead — hydrogen is a zero-carbon fuel.

Although some hydrogen may be generated by electrolysis from renewable fuels, it is likely that much of it will be produced from fossil fuels such as coal and natural gas, with carbon capture and sequestration part of the process (Figure 34). Hydrogen can be cost-effectively stored so it could be used as a substitute for oil in fuelling transport.

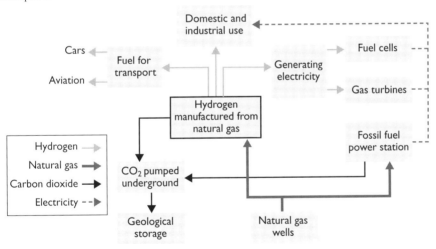

Figure 34 Carbon capture

Can fossil fuel use be made sustainable, i.e. cleaner to use and incorporating improved technologies? The answer is maybe. In terms of futurity, even if fossil fuels are used more cleanly they will still be subject to depletion, maybe at even faster rates. In terms of environmental pollution and achieving environmental sustainability, the clean-up can certainly cut carbon dioxide emissions, but by how much? Are renewables potentially better? In terms of sociocultural sustainability, cleaner fossil fuels will certainly improve health and quality of life for the world's people.

However, geopolitical issues remain with oil and to an extent natural gas, which can threaten economic sustainability. Even some HEP schemes can release methane, which is an even more powerful greenhouse gas than carbon dioxide.

First-generation **biofuels** (including biodiesel made from palm oil and soya beans, and bioethanol made from corn and sugar cane) are marketed as environmentally friendly clean fuels and lay claims to be sustainable. However:

- When plant-derived fuel, be it biodiesel or ethanol, is burned in an engine the carbon dioxide released is offset by the amount of carbon dioxide absorbed by the plants, making it theoretically carbon neutral.
- Unfortunately, the energy needed to plant, tend, harvest, process and transport the finished product makes the above equation less favourable.
- To grow biofuels, land is used that could be used to grow cereals to feed the growing numbers of starving people in the world, so adding to the food crises.
- To grow biofuels such as palm oil in Indonesia or soya beans in Brazil often requires the removal of high eco-value tropical rainforest, which leads to biodiversity loss.

The debates surrounding biofuels versus food and biofuels versus biodiversity mean that the sustainability of biofuels is questionable. The main motivation for the USA's dash for biofuel is undoubtedly energy security. The US government heavily subsidises the growing costs.

The development of second-generation (brushwood and trees) and third-generation biofuels (algae) may be more sustainable, but the processing technology is complex and they are not likely to be economically competitive with fossil fuels in the foreseeable future.

The sustainability of **nuclear fuel** as an environmentally friendly clean fuel is also highly contentious.

- Nuclear power stations produce high-level radioactive waste in the form of used fuel rods, which have to be removed from reactors — in the UK at the Thorp reprocessing plant at Sellafield in Cumbria. Nuclear waste takes a long time (millions of years) to lose its radioactivity, so where to store it and how to transport it safely to storage sites are key technical and political issues.
- Chernobyl (1986) and the earlier Three Mile Island accident highlight the safety issues from nuclear power. New designs are safer, but there are inevitably concerns about radioactive leaks and possible leukaemia and cancer clusters.

However, uranium is relatively cheap to mine and most commentators think that reserves are plentiful and will last for at least 150 years. The raw materials are used in such small quantities that they are cost effective to transport. In time, fusion will replace fission, making nuclear power partially renewable.

Nuclear power is seen as the fuel to bridge the upcoming energy gapmainly because it will not contribute to greenhouse gas emissions.

Distribution of nuclear power is shown in Figure 35.

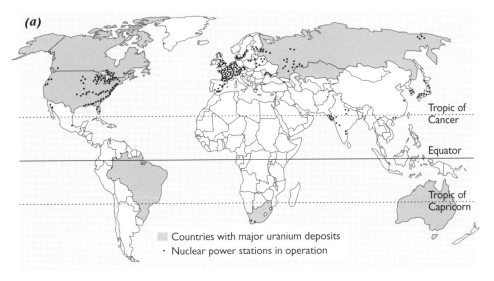

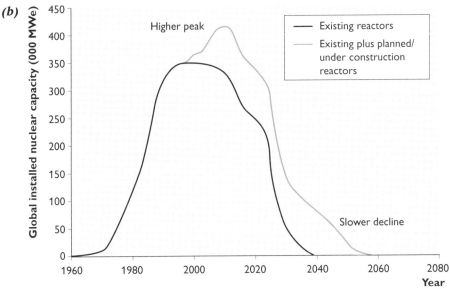

Figure 35 Nuclear power distribution: (a) reactor locations around the world, (b) current and planned reactors

Switching to renewables

Figure 36 on page 72 shows the potential for primary energy generation of the various forms of renewable energy by source. It confirms the importance of solar power — available almost anywhere on Earth. Its potential is however reduced by cloud cover and latitude (Arctic/Antarctic seasonal insolation patterns) as well as wind power.

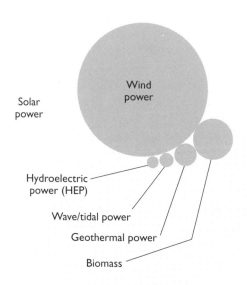

Figure 36 The relative size of total global renewable energy potential by source

Given the necessary techno-fix most renewables have the potential to be eco-friendly as they do not give off greenhouse gas emissions. Most people accept that global warming and its impacts is the most important environmental problem and therefore a move towards renewables is inevitable.

As demand grows, with a rising population living more affluent lifestyles, new, non-polluting sources will be needed to bridge the energy gap and in greater quantities as part of a multi-energy solution. Figure 37 shows the factors contributing to the uncertainty over global energy supplies.

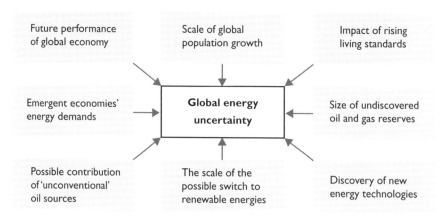

Figure 37 Factors contributing to the uncertainty over global energy supply

As a multi-million-pound high-tech fix is required for tidal, wave and solar power to make them commercially available globally, poor communities in remote rural areas are much better served by a combination of micro-hydros, biogas converters, mini wind turbines and solar thermal systems — collectively known as **renewable microgeneration**. These highly sustainable, small-scale, community-based, bottom-up projects use relatively high technology, so they are especially widespread in Germany (1 million microgeneration units installed as they are government-grant aided) and other developed nations. However, they are increasingly playing an important part in developing nations such as India, Pakistan and Nepal.

In **India**, barefoot solar engineers are trained to assemble, install and maintain solar technology in the rural villages of Rajasthan. They serve 100,000 homes in over 100 villages. In **Pakistan**, the Alternative Energy Development Board (AEDB) has launched solar home systems to provide villagers with lighting, cooking and water disinfection. Each household has a kit that comprises an 80 W solar panel, a charge controller, a battery, four compact fluorescent lamps, two LED lights, a 12 V DC fan and a television socket! This is an example of **leap frogging** as the electricity is provided to villages and communities outside the reach of grid-based suppliers. In **Nepal**, the widespread use of community-based micro-hydros is bringing the same comforts to the remote Himalayan villages.

Therefore, the conclusion has to be 'bring on the renewables'. Even with a number of disadvantages of economic cost and some environmental non-greenhouse gas issues, they are definitely more sustainable alternatives to fossil fuels.

Using energy more efficiently

Energy conservation refers to the variety of methods by which the use of all types of energy, especially electricity and motor vehicle fuel consumption, is limited by efficiencies or reduced by conservation. This may be achieved by:

- greater efficiency, for example more economic fuel consumption and cuts in emissions in cars, or savings in the home with the use of cavity wall and roof insulation and low-energy light bulbs, which cut down wastage
- the switch to energy sources that are less wasteful or renewable (for example, solar panels)
- devising alternative strategies (for example, teleworking and web conferencing to avoid unnecessary travel)

The initiatives available in the UK include both inducements (carrots) and deterrents (sticks). Initiatives include:

- The introduction of smart meters, which show people how much energy they are using, thereby encouraging them to change their usage patterns in order to make savings.
- Tougher environmental standards for new cars and homes.
- Working with industries to develop energy-efficient A* rated appliances such as tumble dryers, as well as encouraging people to scrap energy-inefficient goods (for example, the car and boiler scrappage schemes of 2009).

- Public awareness campaigns such as 'kill your standby' or be green and ride your bike or walk.
- Setting up carbon-trading schemes for companies to cut greenhouse gases.
- Providing high quality, affordable public transport to cut car usage.
- Providing grants for public procurement of low-carbon vehicles such as hydrogen-powered buses or electric cars, or for renewable energy installations in villages and private houses.
- Offering buyback schemes for any surplus energy produced by these renewables.
- Developing taxing strategies — so-called green taxes — that favour, for example, those using energy-efficient vehicles.

A combination of the above strategies will undoubtedly begin to slow demand, conserving resources for future generations and reducing environmental impacts — a key facet of energy sustainability.

3.5 Can a sustainable energy supply be maintained in the future?

The penalty of failure is enormous: Armageddon in terms of environmental damage, and supplies of oil, gas and fuelwood running out. It is oil that concerns the most people as they can barely cope with a one-day petrol strike without panic buying and hoarding.

However, alternatives to fossil fuels are being developed at an ever-increasing rate, but no single energy service is the answer. Both globally and nationally, a diverse mix using a combination of resources, together with real demand reduction by efficiency savings and conservation, is the only truly sustainable way ahead.

Theme 4: Sustainable cities

4.1 The issues

This theme overarches many of the issues that you have studied in the previous three themes. Cities, and indeed all urban areas, need feeding, cannot survive without water and consume energy in order to function.

The Sustainable Cities Programme (SCP) is a concept established jointly by UN-HABITAT and the United Nations Environment Programme (UNEP) in the early 1990s. Sustainability for cities is based on three underpinning forces:

- environmentally sustainable urbanisation

- economic sustainability of urban areas
- social justice leading to equal access to facilities and equal quality of services

Addressing these forces and the consequences of the development of cities over the centuries is the role of the regulatory framework and urban planning. The move from expert-led technocratic planning to stakeholder and community-led planning is seen as the underpinning foundation for the creation of sustainable cities. Girardet defined a sustainable city as 'a city that works so that all its citizens are able to meet their own needs without endangering the wellbeing of the natural world or the living conditions of other people, now or in the future'.

Figure 38 applies the concept of the sustainability quadrant with which you are now familiar to the issue of sustainable cities.

Futurity	Environment
Cities cater for the present on the basis of past technologies, which also have to provide for future generations and new technologies.	The activities of cities consume energy, water and food, and produce waste and effects on the environment. They modify the environment in which they are located, often to the detriment of future inhabitants. Differing built environments will make different demands on resources to become sustainable.
Public participation	**Equity and social justice**
People are stakeholders in the future of where they live, either through the ballot box or by direct action.	People need equal access to education, jobs and housing including services, water supply and waste disposal. Protection by the law and laws that enable human development are fundamental. Wellbeing is the ultimate goal.

Figure 38 The sustainable city quadrant

4.2 How can cities throughout the world be classified?

You have already come across some ideas on the classification of cities in Unit G2. Cities can be classified by size, rate of growth and level of development.

Size
- Towns are urban areas with a range of facilities such as post offices, schools, and mainly independent shops.
- Cities are large settlements depending on commerce, manufacturing and service industries.
- Conurbations are urban areas that have gradually fused together yet maintain different centres.

- Metaconurbations are bands of contiguous urbanisation, first proposed by Yona Friedman in the 1960s as a merging of conurbations whose sheer size is very extensive. The European Banana — the urbanised heartland of Europe stretching from northwest UK to London; through the highly urbanised Nord region of France, the Netherlands and Belgium; the Rhine valley; Paris; the Rhone valley; to the Po valley in northern Italy and Mediterranean France — is often seen as a potential metaconurbation.
- Megacities are those cities with over 10 million people that are centres of the global economy. Today in 2010 there are three megacities — London (11.9 million), New York (21.8 million) and Tokyo (34.1 million). In your lifetime Shanghai (17.9 million) may achieve global hub status. Other very large cities are not global hubs.
- Millionaire cities are cities whose population is a million or more, but is a rather dated concept. Metacities is a more modern term and refers to the largest cities such as Mexico City (22.6 million), Seoul (22.2 million), São Paulo (20.2 million), Mumbai (19.7 million), New Delhi (19.5 million), Los Angeles (17.9 million), Jakarta (17.1 million) and Osaka (16.8 million).

Other ways of classifying by size include:
- size of the population employed in tertiary employment
- the areal extent of the city
- the extent of redeveloped or new areas to accommodate business — this could include the emphasis on vertical extent
- the size of the younger age groups living in the city — the future citizens
- the number of private vehicles
- the size and relative importance in employment terms of industry, government, cultural activities, retailing, education, administration and financial services

Rate of growth

Urbanisation is the process by which the proportion of the population living in an urban area increases. It is a process of change that affects both the places themselves and the people involved. It is also a multi-strand process whose characteristics vary over both time and space. In particular, urbanisation in MDCs took place alongside industrialisation in the nineteenth century and the growth of the tertiary and quaternary sectors in the late twentieth century. In contrast, urbanisation in NICs and LDCs has taken place without industrialisation as cities became centres for services and government. In 1900 only 5% of the world's population lived in cities and in 2008 over half lived in cities. The urban population is projected to reach 6.4 billion or 70% of global population by 2050. The distribution of increasing urban population will not be even, as Table 11 shows for 1950–2050. In MDCs urbanisation reached 50% in the 1950s whereas the developing countries will not achieve that level until 2020.

Table 11 Global urban population, 1950–2050

Region	Urban population (millions)				% urban			
	1950	2007	2025	2050	1950	2007	2025	2050
World	737	3,294	4,584	6,398	29.1	49.4	57.2	69.6
More developed	427	916	99	1,071	52.5	74.4	79.0	86.0
Less developed	310	2,382	3,590	5,327	18.0	43.8	53.2	67.0
Africa	32	373	658	1,233	14.5	38.7	47.2	61.8
Asia	237	1,645	2,440	3,486	16.8	40.8	51.1	66.2
Europe	281	528	545	557	51.2	72.2	76.2	83.8
Latin America and Caribbean	69	448	575	683	41.4	78.3	83.5	88.7
North America	110	275	365	402	63.9	81.3	85.7	90.2
Oceania	8	24	27	31	62.0	70.5	71.9	76.4

Source: United Nations, 2008

In the developed world city populations are increasing slowly (0.8% per annum between 1975 and 2007) although there are cities, especially in eastern Europe, that lost population as a result of migration to the EU, for example Sofia, Lodz and Budapest. In North America regional migration accounted for changing urban populations; in the USA the cities of the southwest (Phoenix, Houston, Dallas) grew while the industrial northeastern cities such as St Louis (–59% between 1970 and 2000), Pittsburg, Cleveland and Detroit all lost about half their populations.

Asia is urbanising rapidly, although the pace is forecast to slow down when many cities become wealthier as the RICs (India and China) develop, for example Shanghai, Shenzen, Mumbai, New Delhi, Jakarta and Seoul. Sub-Saharan Africa is urbanising rapidly, as people are moving to cities despite economic stagnation. The UN describes this as 'urbanisation of poverty in the form of informal settlements'. About 200,000 people migrate to cities across the world every month, which is about the population of Santiago or equivalent to almost the population of Swansea (229,000).

Level of development

Urbanisation is commonly thought to reduce poverty, homelessness, social exclusion, segregation and crime and increase equality. However, even in highly urbanised Western Europe in recent years, contracting economic growth has increased inequalities in cities in the UK, although Western Europe is more egalitarian than other continents. The **transition countries** (those emerging from a centrally planned economy) have increasing levels of inequality. Urbanisation in developing countries is causing increased levels of poverty in cities, leading to more slums. One-quarter of households in the developing world live in poverty and this rises to 40% in African cities. Between 25% and 50% live in what the UN calls **informal settlements** (for example, shanties, favelas and bidonvilles). Fewer than 35% of the people in these cities have their waste water treated, so it is no surprise that half the urban population suffer from diseases associated with inadequate water and waste water disposal. Some 1.2 billion people globally (approximately the population of China) have no access to clean water.

The **Gini index** measures the extent to which the distribution of income among people deviates from a perfectly equal distribution. An index of 1 is absolute inequality whereas 0 signifies perfect equality. It can be measured for countries (Table 12) or cities. National data often mask greater variation in cities and is not always for the same year, as Figure 39 shows for African cities.

Table 12 Gini index for a selection of countries

Gini index	Exemplar countries	Gini index	Exemplar countries
0.25	Czech Republic, Denmark, Sweden	0.40	Russian Federation
0.26	Norway, Bosnia and Herzegovina	0.41	USA, Ghana
0.32	Canada, South Korea	0.47	China, Nepal, Rwanda
0.34	UK, Indonesia, Vietnam, Greece	0.57	Brazil, South Africa
0.37	India, Uzbekistan	0.63	Sierra Leone, Lesotho

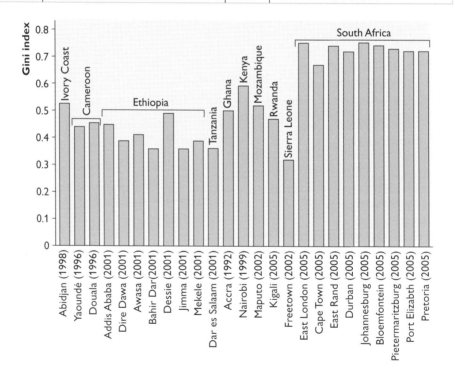

Figure 39 Gini index for selected African cities — why is inequality greater in South Africa than in Ethiopia?

4.3 What pressures currently confront cities and how are they changing them?

The urban challenges of this century can be grouped as follows:
- Environmental challenges, including:
    - climate change (for example, the Stern Report of 2006, which you may have studied as a part of Unit G1)
    - the depletion of oil supplies and the rising cost of fuel (see Theme 3 and **www.sustainablecities.org.uk/energy**)
    - the environmental risks posed by the location of many cities, especially in LDCs in disaster-prone regions
- Economic challenges, such as the effects of financial crises.
- Changes in governance to encourage people to manage their own cities.
- Changes in urban society as a result of migration and natural increase, together with increasing disparities in wealth between the richest and poorest inhabitants.
- The physical sprawl of cities.

The pressures vary depending on the level of development of a country.

The box below provides basic data on Chennai, India and is adapted from the WJEC World Development (WD3) Resource Folder, February 2010. Throughout this section data from the same source are used to provide a single city summary.

Box 1: Introduction to Chennai

Chennai is the fourth largest city in India and is the capital of Tamil Nadu state. It is a focus of economic, social and cultural development in southeast India. The urban area is growing very rapidly and there is considerable pressure on space, resources, basic facilities and services. The whole metropolitan area consists of the city itself plus a large area of rural villages and small towns. The city landscape is a mix of industries, dilapidated slums, modern apartments and wealthy suburban developments. Chennai is part of the Sustainable Cities Programme funded by the United Nations Development Programme to deal with its economic, social and environmental problems.

	Population (millions)		
	1981	**2001**	**2021**
Chennai city	3.28	4.34	5.54
Chennai metropolitan area	4.60	7.04	11.19

- Over 1 million people live in slums.
- 45,000 people are homeless.
- 5% have no access to a toilet in the home.

- 23% of population are illiterate (2001 census).
- There are 75,000 child beggars.
- 72% are lowest caste (Dalits).
- It has the highest crime rate against women in India.
- There is a 50% per year increase in cars.
- There is a 65% per year increase in bicycles.
- 55% workers are in the informal sector.
- 87% households live in permanent houses.

Source: after WJEC World Development (WD3) Resource Folder, February 2010

Transport

The relationship between low-density cities and the excessive use of energy was officially recognised at the Rio conference in 1992. In the Seattle region, weekly travel miles have increased from 60 per person in 1990 to 82 per person in 2007. Such increases are not sustainable.

In Atlanta the density of population is six persons per hectare and only 4% of the population are close to the metro, resulting in only 4.5% of trips being made by public transport. In contrast Barcelona, with a density of 171 people per hectare and 60% of the people within 600 m of the metro, has a resultant 30% of trips made by public transport. Accidents kill 1.2 million people globally each year. Mixed land use rather than over-large land-use zones provide the best opportunity to increase environmentally friendly commuting and reduce car dependency.

The pressures are on to make transport in cities less carbon intensive. If there is a shift to lower carbon forms of movement such as walking, cycling and public transport then improved air quality, better health, improved quality of life, less congestion and, eventually, more attractive, competitive towns should result. Much will depend on location of activities and planning safe, high quality environments that use, for example, 'green corridors' to link homes, schools and services. A 20% increase in cycling by 2015 would save money by improving the health of the population and the costs of unhealthy car-based lifestyles on health provision. A **cycle and ride scheme** links the city-wide cycle routes to the rail and tram network in Copenhagen. In Paris **Velib**, a system of cycle hire from pick-up and drop-off points, is being copied in other French cities. **Traffic cells** are areas whose borders cannot be crossed by cars but only by public transport, bicycle or foot and this idea has been successful in Bologna in Italy.

In the UK, light rail systems (trams) such as the Tyne and Wear Metro and the Manchester tram lines have a part to play in encouraging regeneration of areas and reducing car usage. Bus priority lanes and guided bus systems, such as those in Leeds, are attempts to encourage public transport use.

Land

Cities cover 25% of the world's land surface yet they use 75.5% of the world's resources. The ecological footprint of cities is vast. London's footprint is 20 million hectares compared with its area of 159,000 hectares — 125 times its area. Each Londoner requires 1.2 hectares for food production. In the Seattle region, the built-up area now covers 20% of the land while the proportion of woodland and agricultural land has declined by 5% in the past 15 years. In the developed world, the loss of land to urban activities continues — land for retail parks, airports, suburbs, golf courses and sports stadia.

Housing

Sustainable energy planning and management reduces energy consumption and may increase reliance on renewable energy sources. The design and construction of buildings is important because over 50% of energy use is consumed in buildings. Some of the benefits of sustainable energy use in cities are global — it should lead to healthier, less polluted environments both in and beyond the city. Much of this must come from local governmental initiatives to cut consumption by, for example, subsidising the retrofitting of buildings with insulation and energy-saving lights, utilising new forms of street lighting and subsidising the costs of converting boilers (UK boiler scrappage scheme). Copenhagen has developed a series of district heating schemes that use less energy. In Barcelona it is now compulsory to install solar collectors for hot water in new and refurbished buildings.

One in three people in developing countries live in slums (62% in sub-Saharan Africa and 43% in south Asia), which are frequently built either on the periphery or on land that is liable to landslides and flooding. These **informal settlements** are often the product of people moving on to land and squatting there, as is common around Caracas in Venezuela. In Brazil the Roofless Workers of the Centre hands out undeveloped land and empty buildings for occupation.

Crime

Crime is an often undeserved relation to urbanisation, but much depends on the timescale and the data used. For example, recorded crimes in most UK cities declined between 2008 and 2009. The areas with the highest crime rates per 1,000 people include Islington and Newham in London, Nottingham, Leicester, Cardiff and Southampton, yet all of these cities had declining absolute numbers of crimes. Some of the towns with the highest absolute increases were in leafy suburbs such as Woking and Runnymede.

Some of the solutions to urban overcrowding, such as underground railways, are themselves the focus of horrendous crimes, for example the 7/7 bombings in London and the sarin attacks on the Tokyo subway. Fear of crime is leading to gated communities in South Africa, the USA and Latin America.

Water supply

This topic has much in common with Theme 2. Bristol used 56,000 litres per household in 2007–08, 18% of which was as a result of leakages. UN–HABITAT estimates that up to 40% of water supply is being lost due to the poor state of the system, especially where it was installed over a century ago. Sustainable water management also involves the management of flood risks to prevent damage to the existing stock of buildings.

A lack of water in homes in LDCs increases health risks. It is also burdensome for women, who have to fetch water, and it prevents good hygiene practices.

Box 2: Water in Chennai

There is an acute water shortage in Chennai. Rainfall is unpredictable and droughts are common. Underground freshwater reserves are being pumped out faster than water is being added from rainfall. Many wells and boreholes have been dug in villages around Chennai. Over 6,000 tanker loads a day supply the city. At least 4,000 of these supply multi-storey apartments, hotels, hospitals and businesses. The government has been ineffective in taking action on water, so the tanker business is lucrative. Over 200 bottled water companies make large profits. There are no licence fees for water extraction and businesses ignore any restrictions on exploiting water.

The urban expansion around Chennai creates severe problems in providing amenities such as drinking water, sanitation, solid waste and sewage management. The local councils suffer from lack of financial resources and support from the city authorities and there is a need for public education about water conservation techniques such as rainwater harvesting.

Source: after WJEC World Development (WD3) Resource Folder, February 2010

Resource use and waste disposal

In meeting future needs, cities must minimise the use and waste of non-renewable, finite resources such as energy. In the UK only 50% of the 120 million tonnes of building waste is recycled. Cities possess cultural and historic assets that are non-renewable, so these must not be wasted. A city's ecological footprint must be managed so that, for example, its florists do not depend on roses grown in Kenya. Biodegradable wastes must not affect ecosystems and cause them to degrade. Rivers have a capacity for biodegradation that should not be exceeded. Non-biodegradable waste and emissions must not overburden the capacity of landfill sites and global sinks so that they can no longer cope and ecological changes result. Waste can be used to provide energy rather than being put into landfill.

In LDCS the most severe environmental hazards are often found near people's homes. Shared toilets lead to diseases such as cholera and other diarrhoeal diseases. If there is no rubbish collection, waste accumulates and flies and rats spread disease. In Accra, Ghana, 45% of households have no water source in the home and a similar percentage share toilets with up to 10 other households. Over 80% of households still lack rubbish collection. The use of wood, paraffin and candles for heat and light causes much pollution and is costly.

To the east of Kolkata in India is a waste-water fed agricultural system. Figure 40 shows the main features of this scheme. It provides a livelihood for 17,000 residents who depend on this natural waste water treatment plant for the fish ponds, water for paddy fields and decayed vegetable matter for fertilising vegetable crops.

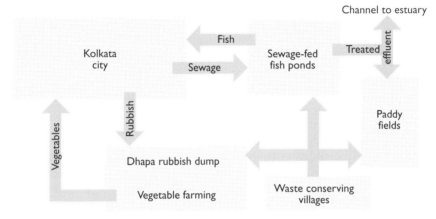

Figure 40 Waste-water fed agricultural system in the Kolkata wetlands

In Manila, Philippines, 98% of the people have access to clean water yet only 25% of those households have access to sanitation, while in Jakarta the figures are 70% and 1.5% respectively. Table 13 provides data on access to piped water and sewage for a selection of cities.

Table 13 Access to piped water and sewage

City, country	Households with access to piped water (%)		Households with access to sewerage (%)	
	1990	2003	1990	2003
Mumbai, India	68.0	82.2	30.3	41.1
Kolkata, India	30.5	37.9	42.7	44.0
Beijing, China	ND	97.7	ND	47.6
Shanghai, China	ND	99.7	ND	66.2
Kampala, Uganda	11.9	15.1	2.0	13.6
Accra, Ghana	63.3	55.5	12.7	36.6
Rio de Janeiro, Brazil	ND	88.5	ND	63.3
Curitiba, Brazil	ND	84.2	ND	55.4
Bogotà, Columbia	ND	100.0	ND	100.0

Source: after United Nations Centre for Human Settlements, 2009

Environmental quality

Cities are inevitably hotspots for disaster risk according to UN-HABITAT. Of the 10 largest cities in the world, 8 are in earthquake-prone zones, 9 are in areas affected by storms and hurricanes and 8 are in areas of severe flood risk. Six cities are liable to storm surge. Some are prone to multiple hazards, for example Mumbai and Shanghai are liable to earthquake, storms, flood and storm surge, and Tokyo to earthquakes, storms, tornados, floods and storm surge. Disasters continue to affect all sizes of urban areas, for example the Workington and Cockermouth floods in November 2009, the 2005 hurricane in New Orleans which killed 1,800, the 2003 heat wave in Paris which was estimated to have caused 14,800 deaths (not only in Paris but across France) and the Bam (Iran) earthquake in 2001 which killed 19,700.

In Chennai traffic congestion, air pollution from traffic, overcrowding, and the impact of expansion on already stretched sanitation, waste and sewage management are severe. In Accra over half of households still use wood or charcoal for cooking, which reduces air quality in the home and leads to respiratory diseases.

The consequence is that our culture has developed a fear of cities because they are 'super suburban' — a spread of suburbs that are of the city yet are the areas that people escape to.

In summary, how have attempts to control city spread in the UK succeeded or failed since people realised the need to control city growth a century ago? Table 14 summarises these changes.

Table 14 Phases of city planning that indirectly attempt to create sustainability

Means of controlling urban form	Approximate date	What it achieved	Evaluation
Garden cities	Around 1900	Small towns, low densities, large plots, functional zoning	Small scale, too utopian
Suburbia	1920 onwards	Free market expansion, medium- and low-density residential areas linked to improved transport	Problematic in the age of the car
Green belts	1937	Constraint to the spread of cities by restricting development	Successful but development leapfrogged the green belt
Neighbourhood units	1940s	Attempt to create a social unit focused on community facilities. Based initially on the catchment area of primary schools	Not all modern community facilities can be provided in an area

Modernism	1950s	Le Corbusier's ideas transformed to provide high-rise dwellings and industrial building	Ronan Point Disaster of 1968 led to concerns about safety. Social issues led to the demise of building high-rise blocks except tower blocks for affluent apartments
Renewal/slum clearance	1950s	Has taken many forms – high-rise and high-density, low-rise. Peripheral social housing	Can result in areas of deprivation being hidden
Road hierarchies	1960s	Buchanan Report	Too much emphasis on roads at the expense of other modes of transport
New Towns	1945	Often using ideas from initiatives listed above	Product of specific time periods

4.4 What attempts have been made to find sustainable solutions to problems faced by cities?

Global scale

(1) Urban Management Programme of UN–HABITAT (1986). This programme aims to strengthen the contribution of cities in LEDCs to economic and social development and it has focused recently on the urban poor. One hundred and forty cities in 58 countries are involved.

(2) Sustainable Cities Programme of UN–HABITAT (1992). This is a joint programme of UN–HABITAT and the United Nations Environment Programme (UNEP). Chennai is part of the Sustainable Cities Programme funded by the United Nations Development Programme (UNDP), which has helped the city to deal with its major economic, social and environmental problems. Figure 41 shows the main elements and priorities of the programme.

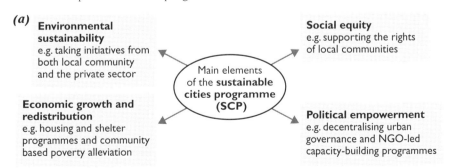

(a)

Environmental sustainability
e.g. taking initiatives from both local community and the private sector

Social equity
e.g. supporting the rights of local communities

Main elements of the **sustainable cities programme (SCP)**

Economic growth and redistribution
e.g. housing and shelter programmes and community based poverty alleviation

Political empowerment
e.g. decentralising urban governance and NGO-led capacity-building programmes

(b)

Source: after WJEC World Development (WD3) Resource Folder, February 2010

Figure 41 Sustainable Cities Programme: (a) main elements, (b) priorities

(3) Healthy Cities Programme of WHO (1986). This programme was established to improve and promote health in urban areas by bringing together all of the agencies responsible for health. It has already been noted in Glasgow that elderly people living close to parks are healthier than those who cannot reach them.

(4) Safer Cities Programme of United Nations Centre for Human Settlements (UNCHS) (1996). The Manzese area of Dar es Salaam (Tanzania) accounted for 25% of all crime in the city and especially crimes against women. A safety audit was carried out by women, which resulted in a series of proposals to make the area safer by improving lighting, widening streets, using areas between houses for allotments and play areas, and improving drainage and sewage disposal.

(5) Local Agenda 21 (1992). This grew out of Agenda 21 of the United Nations Conference on Environment and Development (UNCED), particularly the chapters relevant to sustainable cities.

Continental scale

European Green Capital is an initiative in which one European capital is selected each year. The general aim is to improve the European urban living environment and thus the environment as a whole. Stockholm won the title in 2010 and Hamburg in 2011. It is given to a city that:

- has a consistent record of achieving high environmental standards
- is committed to ongoing and ambitious goals for further environmental improvement and sustainable development
- can act as a role model to inspire other cities and promote best practices to all other European cities

The cities will act to solve environmental problems and improve the quality of life of their inhabitants and to reduce their impact on the global environment. To qualify, a city must have in excess of 200,000 inhabitants. Other finalists for 2010 and 2011 included Amsterdam, Bristol, Copenhagen, Freiburg, Münster and Oslo.

National scale

Compact Cities

Curitiba (Brazil) pioneered the compact city approach to lowering the time and cost of travelling in 1965. The city (1.6 million people) has a set of transport axes for buses, which are bordered by commercial centres and high-density residences. Stops on the bus routes are the focus for links to other local transport. The effects are:

- 75% of commuters use public transport
- residents spend 10% of income on transport
- traffic levels have declined by 30% since 1974
- fuel consumption is 25% lower than other Brazilian cities
- 85% of residents (1.3 million people) use the bus each day
- it is self-sufficient financially

However, the scheme is not a total success. Many of the poor live on the outskirts of the area beyond the public transport system. The houses near the transport nodes and along the axes have risen in price and altered the social structure of these areas to a more middle-class population.

The integration of transport and land use pioneered by Curitiba has been copied elsewhere in South America, for example in Bogotá's Transmilenio.

Integrating rail and bus transport is a simple solution by creating transport hubs at railway stations (as in Naples) and following this up by integrating timetables. A similar project, called **concentrated deconcentration**, was started in the 1960s in Amsterdam. The urban area has been made accessible for all by creating a polycentric set of hubs beyond the city centre, all linked by a range of transport modes, including bicycles, designed to keep the city compact. A measure of its success is that 35% of commuting journeys are made on foot or bicycle.

Smart cities

The smart city movement began in the high-tech sector of the economy in the USA. It aims to get cities to use the new technologies to decrease costs of transport, administration, energy and water treatment, among others. A smart city performs well on six characteristics:

- smart economy
- smart mobility
- smart governance
- smart environment
- smart living
- smart people

These six characteristics comprise 30 indicators, which are aggregated for each characteristic.

Informal networks (Bilbao), amalgamation of districts into a greater whole (Montreal) and cross-border international city collaboration (Basel–Mulhouse–Freiburg Regio

Basiliensis and Copenhagen–Malmo) are cities using technologies to manage their operation and development.

Cardiff and Portsmouth are two of the UK's smart cities. Their smartness has been aggregated in Figure 42 (scores above 0 are good, scores below 0 are not up to standard). The scores for mobility, environment and living are subdivided by factors in Figures 43 to 45. Go to **www.smart-cities.eu** for links to all the European smart cities.

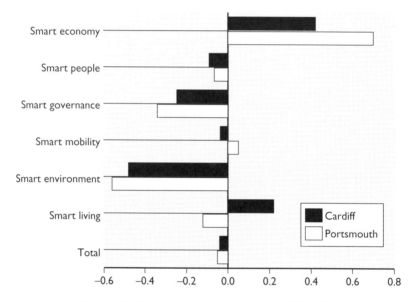

Figure 42 Smartness aggregate: Cardiff and Portsmouth

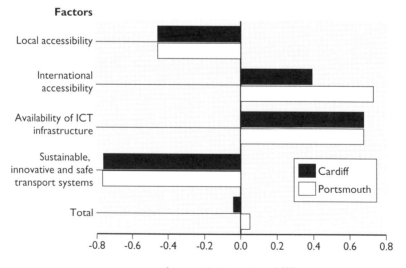

Figure 43 Smart mobility

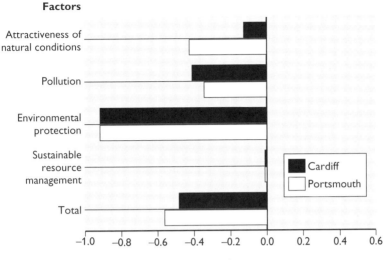

Figure 44 Smart environment

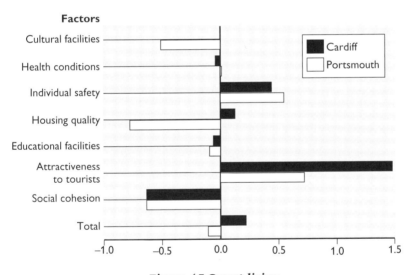

Figure 45 Smart living

Smart technologies have been developed to predict traffic flows in the Singapore CBD over periods up to an hour with 85% accuracy, based on data collected within the congestion zone and beyond. Smarter city initiatives in New York enabled a 27% drop in crime since 2001.

New Towns and Garden Cities

The Garden City movement established by Ebenezer Howard in the early twentieth century attempted to recreate the English village by building manageable towns in which the countryside came into the town. Howard believed that a population of 32,000 in a municipally developed town was the ideal population size for community life. Letchworth was the first town to be developed along Garden City principles. In 1919 Welwyn Garden City was begun, which subsequently became one of the first generation of New Towns, a scheme that adapted Howard's principles to the postwar world.

New Towns in the UK tended to follow the planning dogma of the time and were not necessarily focused on sustainable urbanisation. For example, the road network in Milton Keynes was constructed for the car. However, as the towns have evolved some aspects of sustainability have been adopted. Crawley has established segregated bus routes that are similar to those in Almere in the Netherlands.

Redevelopment and reinvention
There have been four approaches to achieve sustainable urban development in already built urban areas:
- Self-reliant cities are based on the bioregion needed to supply a city rather than a political unit. This idea is sometimes called Ecotopia because it uses utopian ideals rather than reality.
- Redesigned cities are created around a more traditional approach that has been adopted in an attempt to make the city more sustainable by cutting sprawl, building public transport infrastructure and cutting journey distances and times.
- Externally dependent cities are those that are based on the premise that economic growth will enable people to solve the problems posed by cities. It is a market-orientated approach that is called into question by economic recession and lack of funds/taxes to support change.
- Fair-share cities are an attempt to trade the costs of a city on the environment so that the inhabitants pay for the benefits they gain by covering the environmental costs of the surrounding areas.

Transition Towns
Transition Towns are a community-led response to the pressures of climate change, fossil fuel depletion and economic contraction. They are the result of bottom-up organisations that start from small group initiatives which address food supply, transport, energy, housing and education and create practical projects. Other developments include car clubs, local currencies, neighbourhood carbon-reduction clubs and urban orchards. Totnes in Devon was one of the first towns to embrace the idea (**http://totnes.transitionnetwork.org**). At a micro scale there is the application of the concepts to transition streets (**www.transitionstreets.org.uk**).

Eco-towns
These began in 2007 as a proposal for ten new communities, which had been reduced to four in 2010. They are Rackheath, Norfolk; northwest Bicester, Oxfordshire; Whitehill Bordon, east Hampshire; and the China Clay Community near St Austell, Cornwall. Figure 46 shows the location of the towns and the other proposals.

The homes will have smart meters to track energy use, community heat sources and charging points for electric cars. Affordable housing (30%) will use renewable energy sources and sell surplus energy to the grid. Parks, playgrounds and gardens are

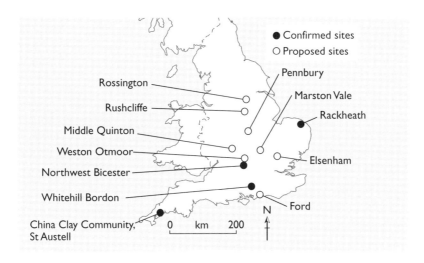

Figure 46 Britain's new eco-towns

planned to cover 40% of the area and there will be strict rules on public transport, with all homes located within ten minutes' walk of bus or train services.

Critics suggest that developing greenfield sites is not eco-friendly. However, eco-towns make use of underused sites, for example the Bordon military ranges and the former air station at Ford. Go to **http://ecotownsyoursay.direct.gov.uk** for further information.

Cities for Life

The Peruvian Cities for Life Forum involves 41 organisations including 21 cities and five universities in a country where 75% of the population live in cities. Just over one-third live in slums and 43% live below the international poverty level of US$1 per day. Poverty prevents environmental improvements because of in-migration, insufficient housing and services and the lack of political will to change. Exemplar best practice was used to encourage further schemes. Among these were:

- Pantanos de Villa — a marshland area and one of the last green areas in Lima; converted to a park with resources for visitors and research
- City of Ilo — incremental environmental management through small-scale projects; bottom-up development
- San Marcos–Cajamarca — improvements to water supply and hygiene to combat cholera
- micro enterprises for the collection of solid waste in Lima
- local non-governmental organisation (NGO) used to train 12,000 people in water management and credit for building sanitary facilities for families

The project is being consolidated in most cities with the aim of promoting sustainable tourism, especially on the coast, building developments that take into account natural hazards and developing programmes to combat the issues posed by air quality, water supply, sewage and waste disposal. The latter aim had no funding in 2004. Universities now provide courses in sustainable development. The website **http://eau.sagepub.com/cgi/reprint/16/2/249** provides greater detail on the project.

Local scale

Ikotoilets is a project started in Kenya to improve cleanliness around communal toilets. The scheme permits small businesses to start up around toilets such as magazine stands, shoe shine stands and booths selling airtime for mobile phones. It is in the interests of the businesses to be in a clean area and the evidence is that there is far less sewage spilling into the vicinity of the Ikotoilets. This type of small-scale project can be scaled up, for example Sulabh International is installing latrines and pay-for-use toilets across India.

The 'walking school bus' is a solution for school-run traffic congestion and its associated waste of energy. There are many examples such as *scuolabus a piedi*, the Rome solution for all primary schools. Cycling to work is a similar solution to cut traffic congestion and energy usage. The problem is that it has to be retrofitted into older cities such as Copenhagen, where half of inner-city commuters cycle to work thanks to bike lanes on every street. Vienna has 1,000 km of cycle paths and has now opened a housing area specifically for cyclists. Bike City, a housing project in Vienna specifically for cyclists, drew 5,000 applications for 99 flats.

Beddington Zero Energy Development (Bed ZED) Project is a carbon-neutral eco-community in the Wallington area of Sutton. Built in 2002, it comprises 82 homes incorporating a new approach to energy conservation and sustainability. The driving force has been the Peabody Trust, a social housing trust that fights poverty in London. Each south-facing house (for solar gain) uses renewable energy from a wood-powered heat-and-power plant. Insulation is made from recyclable materials. The project has attempted to utilise sustainable transport with good public transport links, a cycling network and a car pool system for residents, all of which are legally binding. Sutton is also retrofitting several other properties as part of a government initiative to demonstrate low-energy usage.

The website **www.sustainablecities.org.uk** will provide you with a range of further examples of schemes that address the issues of sustainability in cities.

4.5 How sustainable are cities?

The goals of sustainable urbanisation set by the United Nations Centre for Human Settlements (UNCHS), are listed below.

Environmental sustainability requires that:
- greenhouse gas emissions are reduced and serious climate change mitigation and actions are implemented
- urban sprawl is minimised and more compact cities served by public transport are developed

- non-renewable resources are sensibly used and conserved
- energy use and waste per unit of output is reduced
- waste is recycled or disposed of in a way that does not damage the environment
- the ecological footprint of towns and cities is reduced

Economic sustainability should focus on local economic development to enable the basic conditions for efficient development of the regional economy and include:
- reliable infrastructure and services including water supply, waste management, transport, communications and energy supply
- access to land or premises in appropriate locations with secure tenure
- healthy, educated work force with appropriate skills
- a legal system which ensures competition, accountability and property rights
- appropriate legislative frameworks including minimum standards of health and safety, non-discrimination and the treatment and handling of waste
- the urban informal sector must be supported as it is vital for an urban economy

Social aspects of urbanisation and economic development must not be neglected. These are in the UN-HABITAT agenda and include:
- equal access and fair and equitable provision of services
- social integration by offering opportunities and space to encourage people to interact
- planning that is sensitive to gender and disabilities
- the prevention, reduction and elimination of violence and crime
- social justice, equal access and equal quality and rights of vulnerable groups should be addressed

Sustainable Seattle

Founded in 1991, Sustainable Seattle is one of the world's leading initiatives working towards a set of sustainability goals. The indicators of sustainability that are used have been adopted by the UNCHS since 1996. Its success is based on the fact that it works at a range of levels. It runs competitions in schools asking students to 'Dream a Sound Future' for the Puget Sound area in 2100. This is based on the dictum 'truth comes from the mouths of babes'. It is building a network of sustainability practitioners, the Sustainability Partnership and Resource Commons (SPARC). The partnership sponsors projects and provides a centre to house sustainability workers. The Sustainable Urban Neighbourhoods (SUN) initiative enables people to take action to bring about sustainable futures in their area.

The website **www.b-sustainable.org** is the online mouthpiece for the initiatives in Seattle and King County. Its approach is to analyse what is happening from sets of indicators, explain why they are happening and why they are important for the future. It is then possible to pressurise government to do something about adverse trends.

Chennai

Chennai has had some success in relocating slums, managing solid waste and improving infrastructure as a part of the Sustainable Cities Programme (SCP). USAid, the Asian Development Bank and the World Bank have provided financial backing, but this is small compared with the funds needed. Lack of money is a major stumbling block. It is easily diverted to profitable projects at the expense of the degraded environments that are the homes of the poor. Finance for sustainable development goes through a range of city government departments and becomes fragmented and uncoordinated, so much so that the priority issues for Chennai (see Figure 41) are lost.

Sustainability for cities is based on three underpinning forces:
- environmentally sustainable urbanisation
- economic sustainability of urban areas
- social justice leading to equal access to facilities and equal quality of services

To achieve these aims, cities need to:
- mitigate and adapt to climate change by reducing greenhouse gas emissions, which will require some tough decisions
- increase energy security (Theme 3)
- promote and build sustainable communities and places
- increase employment and economic diversity
- improve health and wellbeing

The best approaches to addressing sustainable cities will be based on understanding the city as a complex, connected system that geographers, among others, study and understand. This involves a cross-disciplinary approach to sustainability that identifies the winning scenarios for the future of the whole of the city. Table 15 looks at the progress that has been made and the hindrances to progress. It is not complete or definitive. Could you add other evidence of progress and hindrances?

Table 15 Sustainable cities: progress and hindrances

Progress	Hindrances
Many cities have indicators of sustainability	All are at different stages of collection, use different measures and different timescales
Indicators show the results of actions and encourage accountability	Often used in short term rather than for long-term goals
Cities throughout the world are engaged and accept the drive to sustainability	Finance and political will when economic climate is poor
Bottom-up, citizen-based initiatives are growing in number	The need for coordination and leadership, which can be hijacked to block progress
The goals of sustainability are wide and encompass all aspects of city life; social, personal, natural and built environments	Too much to encompass and needs to be prioritised. Governments need to take on some aspects that cannot be left to citizens, e.g. planning rules
Some cities have made large strides and become leaders in particular aspects of sustainability	Much depends on political will and that can change. Large corporations may not buy into schemes because profits could be affected

Approaching the Resource Folder

What is the Resource Folder?

Paper copies of the Resource Folder are distributed in November for the January examination and in March for the June examination. The folder can also be downloaded from the WJEC website: **www.wjec.co.uk**.

The folder contains 16–20 pages packed with information, which may be quite daunting when you first open it. There is a useful set of guidelines on page 2 of the folder. It is best to read these guidelines as soon as possible and then again towards the end of your revision to check that you have followed the good advice. Several of the points will be dealt with here in a little more detail.

The resources normally cover at least two of the four themes in the specification. In January 2010 the themes were Sustainable Cities and Sustainable Food; in June 2010 they were Sustainable Energy and Sustainable Water Supply. The four themes can be combined in any way. The booklet contains figures that can take the form of photos, maps, graphs, diagrams or text. These figures are divided into sections, which should form the basis of your exam preparation. *You cannot take the released Resource Folder into the examination.* You will be issued with a fresh one when you receive your question paper.

What to do with the Resource Folder

Your first task is to study the sections to see what they cover. One way to do this is to create a table like Table 16, which has been compiled to show which parts of the specification can be related to each figure. Your tutor will have the Unit G4 specification or you can get a copy from the WJEC website (**www.wjec.co.uk**).

The balance of figures suggests that questions will be focused on key Questions 3.2, 3.3, 3.4 and, less importantly, 3.1. However, there is some emphasis on 2.2 and 2.3 for water supply.

Once you have completed the table, you need to begin to learn what the figures show in relation to the specification. It is essential that you:
- understand what each figure shows
- know where to find information once you are in the examination room

Next, read and study the resources, which should involve more than one session.

Make notes to say what each figure shows and write down any information you do not understand so that you can ask your tutor for assistance. Where there are tables, maps or diagrams, write notes on what they show and what your interpretation of the data might be. You can annotate the Resource Folder itself because this is your

Table 16 Resources v specification grid based on June 2010 pre-release Resource Folder

Figures

	1/2	3/4	5–7	8/9	10/11	12/13	14	15–17	18	19–21	22	23/24	25–27	28/29	30	31/32
Sustainable energy																
3.1 What problems are associated with the supply of energy?		✓	✓	✓			✓	✓				✓ P				
3.2 How and why is the demand for energy changing?						✓	✓									
3.3 How can the demand for energy be managed sustainably?					✓ P			✓	✓	✓ P	✓		✓ P	✓ P	✓	
3.4 Can a sustainable energy supply be maintained in the future?						✓			✓	✓ P	✓	✓ P	✓ P	✓ P	✓	
Sustainable water supply																
2.1 What physical factors determine water supply?																✓
2.2 How do human activities influence water supply?	✓										✓					✓
2.3 How can water supply and demand be managed sustainably?					✓ P						✓				✓	
2.4 Can sustainable water supplies be maintained in the future?															✓	✓

Shaded cells represent text figures; P represents photographs and text.

study copy and a new one will be issued in the examination room. You cannot take the pre-release folder into the examination room and this is why you need to learn where to find information.

Confirm your interpretation of the figures with your tutor in class.

Identify themes that run through the figures. Some of the themes in June 2010 were:
- sources of energy in case studies of countries at different stages of development
- finite and renewable energy resources
- impacts of different energy uses on the environment
- nuclear power
- hydroelectric power
- power and other basic needs for sustainability, for example water and food (link to Theme 1)

Once you have identified the themes, you could draw another table that links the figures to the underlying themes within them. You could then reconstruct your original notes on the Resource Folder to fit the themes. Follow this up by inserting your class notes and reading notes into the same themes. Now, and only now, are you ready to develop your studies further.

The final page of the Resource Folder contains a list of web-based references. Consult these to see if they can provide additional information. (Some sites do change, or indeed disappear, in the time between setting the examination paper and the accompanying Resource Folder and the folder being issued, which is generally 18 months to 2 years.) The guidelines on page 2 of the Resource Folder suggest how to approach the web-based material. They also remind you that there are many sources of information, so do not forget textbooks, journals and the quality press and media.

Do not try to predict questions or learn pre-prepared answers. To do so often leads to answers that do not address the questions set. Just know your data and where to find it and enter the examination room fully prepared. Go through all your notes made and assembled since the Resource Folder was issued at least the day before the examination but preferably several times. Some tutors test you with quizzes asking where information can be found in the folder.

Exam hint
Use your own knowledge as well as the Resource Folder for answers to Section A.

In the examination room

One useful technique at the start of the examination is to draw another table with question numbers across the top and figure numbers down the side. Because you know where all the information can be found in the figures, place a tick in each figure's cell when it could be useful in answering a question. Your grid will look something like this (the ticks are hypothetical in this case):

Figure	Q 1	Q 2	Q 3	Q 4
1		✔		
2		✔		✔
3	✔		✔	
etc.				

This technique of question planning reduces the time for each answer, but it makes sure that you make full use of the resources to support your answers. Alternatively, you could use the contents page of the Resource Folder in the examination room and mark each question with the relevant figures.

Remember that the Resource Folder is primarily for use in Section A. Section B states 'In this section you may use information from your studies for AS and A2 geography as well as from the Resource Folder and your own research'.

Questions
&
Answers

The questions in this section are based on the January 2010 Resource Folder. This is available to download at **www.wjec.co.uk/uploads/publications/9482.pdf**, and will be needed to answer the questions.

Remember that in the exam you must answer all five questions. You may answer the question in Section B first, if you wish. However, the questions in Section A are best answered in order because they develop a theme.

There is guidance on the cover of the examination paper and on page 2 of the Resource Folder. Read this carefully. Now that you are sitting an A2 paper, which is possibly your last assessment in geography at this level, your answers need to be more sophisticated. This is especially so if you are aiming for the new A* grade.

Examiner's comments

In this section each candidate's answer is followed by examiner comments, identified by the ℮ icon. A final examiner summary is also provided, which indicates the level and mark the answer would have gained overall. All comments are designed to stress the strengths and weaknesses of the candidate's answer. The answers illustrate the qualities needed to gain grades for a particular response. The answers contain typical errors such as irrelevancy, lack of focus on the question's wording and lack of case study examples.

Section A

In this section you may use information from the Resource Folder and your own research.

Question 1

Classification of cities

Use information from your own studies and Figure 1 (page 102) from the Resource Folder to explain how cities might be classified. (10 marks)

Time guidance approximately 13 minutes.

Candidate A

Cities may be classified by their standard of living, the annual income for each home or their life expectancy. A developed city would meet all of the human rights, for example access to clean water and the right to eat. If a country has a low percentage of houses with clean water, this often indicates that the standard of living is low, for example only 15% of houses in Jakarta, the capital of Indonesia, have clean water. Without access to clean water the people will not be able to survive, they will starve as food needs water to grow and good nutritious food needs a healthy source to be of good quality. This leads to food insecurity and malnutrition. A lack of clean water can also increase pest and locust infection, spreading disease rapidly and affecting a country's demography and life expectancy. Water is key to agricultural and crop industries, for example in China 1,000 tonnes of water is needed to produce 1 tonne of wheat. If a country cannot export food its L2 economy suffers. If annual income is low and the population high, like in Bangladeshi cities, the city life may be classed as overcrowded and poor, which can suggest slums like those in Rio de Janeiro, Brazil.

The first and third sentences identify three columns in the data. Human rights are identified as a result of the candidate's own studies. Much of the answer digresses into consequences for people, which was not required. It returns to Figure 1 again in the last sentence. This answer will only just gain a D-grade mark because it looks superficially at some possible criteria for classification and does not follow the instructions in the question. The figure provides many other opportunities for classification. Note that question 1 may not refer to a single resource in the booklet. In future papers it might be more wide ranging.

Figure 1

City	Population (millions)	Average annual income ($US)	Houses with clean water (%)	Houses with refuse collection (%)	Cars per 1000 population	Life expectancy (years)	CDI (City Development Index)**
Accra* (Ghana)	3.35	2280	46	60	24	58	46.6
Bangalore (India)	7.35	2648	47	96	130	66	58.0
Dhaka* (Bangladesh)	12.62	219	80	50	7	64	48.4
Havana* (Cuba)	2.33	2249	85	100	32	79	71.0
Jakarta* (Indonesia)	15.14	2843	15	84	68	72	69.2
Lagos (Nigeria)	10.14	1024	65	8	4	49	29.3
Lahore (Pakistan)	7.95	428	84	50	45	65	61.1
Melbourne (Australia)	3.75	16 845	99	100	500	82	95.5
Rio de Janeiro (Brazil)	12.32	5850	95	88	177	74	79.4
Seoul* (South Korea)	23.43	18 970	100	100	340	78	86.0
Singapore* (Singapore)	4.61	26 590	100	100	135	84	94.5
Stockholm* (Sweden)	1.91	25 030	100	100	390	82	97.4

* denotes capital city

** CDI is the measure used by the United Nations Human Settlements Programme (**http://www.unhabitat.org**)

Source: WJEC Resource Folder, January 2010

Candidate B

By looking at the cities in Figure 1, it is possible to classify these cities through a range of methods.

First, one might choose to classify cities by their population. If a population were to have over 1 million people, this city could be called a millionaire city. All the cities in Figure 1 fall into this category. A city with a population over 10 million may be classified as a megacity. Cities of both these kinds have increased in recent years; at the start of the eighteenth century there were not even 100 millionaire cities, but at the end of the twentieth century this figure had risen to nearly 500.

Another way to classify a city would be through whether it is a capital city. For example, Seoul, Jakarta, Dhaka, Accra, Singapore and Stockholm (all in Figure 1) are capital cities. Another method would be to label cities as global cities with populations over 10 million and being, in most cases, capital cities. These are not the only criteria they must meet. In the case of Dhaka and Tokyo, for example, both are capitals, both are megacities but only Tokyo in Japan is a global city. This is because, unlike Dhaka, Tokyo has a reputation of unchallenged prestige in the world. It is recognised as a centre of political, economical, social and technological changes, unlike Dhaka which does share several similarities with Tokyo. For example, it has the 7th largest agglomeration rate (Japan has the first), but it cannot be called a global city for reasons such as only having 50% of houses with refuse collection.

Increasingly, cities are being classified on their sustainability. This looks at a variety of variables used to assess how effectively a city is providing for its population's needs in the least threatening way. In the UK, for example, the top 20 cities in population were ranked by their sustainability: Brighton and Hove ranked 1st. Edinburgh 2nd, London 10th and Liverpool bottom in 20th place.

> ✒ This is a grade-A answer because the candidate recognises from the outset that Figure 1 is the source of inspiration. The answer uses data from the figure and other ideas from studies for Unit G2 and is synoptic. It also links the response to sustainability with some evidence that has been learned. It does not cover all of the possible classifications because that would be impossible in 10–13 minutes. Neither candidate here used the City Development Index as a summary of many variables.

■ ■ ■

Question 2

Food supplies in cities

Explain how disparities in wealth, and other economic factors, can influence food supplies in cities throughout the world. (10 marks)

Time guidance approximately 13 minutes.

Candidate A

Disparities in wealth can influence food supply in cities as in more developed countries the majority of rich people are situated in cities and therefore rely on exports of food from rural areas. This causes a strain on rural areas, which in 2001 had 40% of land cultivated for food produce. As a growing demand for food in cities increases, the rural supply also needs to increase. This causes more land to be needed, which leads to deforestation and the need to irrigate lands. Another economic factor is the rise in prices of the food market which, in conjunction with disparities in wealth, cause an unequal distribution of food. The UN Food and Agricultural Organization says there is enough food available but it is just distributed wrongly. The cities that can afford produce use up 75% of the Earth's resources despite only covering 2% of the surface. A recent study also showed that people in rural areas in MDCs and slums in cities in LDCs simply cannot afford the food. For example, the UK spends 12% of its wage on food whereas in Africa they spend 80% of their income on food, due to rising prices of food, transport costs of the food and also demands for better quality food, for example organic and free range.

🖉 This candidate has fallen into the obvious trap of not making direct reference to the resources and therefore digresses into other aspects of food supply. The answer talks in generalities and hardly refers to cities. It is a weak grade-C response.

Candidate B

Kenya is probably one of the best examples of how food supply is greatly influenced by wealth. In 2006 it had one of the driest seasons in years and thousands of people were malnourished; 27% of children were said to be starving. However, Kenya was still exporting 62,500 metric tonnes of grain to wealthier populations. In a country where most of the people live in poverty, they were unable to afford food. Some 2,000 farmers were forced to move 30,000 cattle into Uganda to avoid losing their herds. The presidential gardens, however, continued to flourish, simply because of the president's huge wealth. However, it is not only environmental factors that can influence food supply.

In Brazil, increasing amounts of grain are now being used to produce biofuels. This has the effect of causing prices to rise as supply is decreased meaning the poor, once again, go without. Globally, it is estimated that 1 in 6 people is hungry often due to reasons of poverty.

Other factors that cause cities to go without food are ethical and environmental concerns. In Britain we import 80% of fruit and one-third of all food. This helps to support a £200 million industry in Africa, which over 100 million people rely on. Since we have become more concerned about the ideas of food miles and seek closer and more local ways of producing food, such as in the green belt around London or in aeroponic and hydroponic vertical farms, as shown in Figure 12, we have begun to reduce the amount of imported food we eat. This means that already vulnerable LDC populations lose income and start to go without food. In the Horn of Africa 5.4 million people are hungry.

As a result of the Haiti earthquake of January 2010, many people are starving even though money and food supplies have been donated to the country. This LDC is unable to provide the infrastructure to allow food supplies to reach the population, meaning once again that the poorest in the world go without food. However, this time it is an environmental disaster that has had an effect but, essentially, it comes down to disparities in wealth.

> The answer only links the topic to cities in the third paragraph, although it implies the occasional link to feeding cities. It tends to stick with the issue of food supplies. To its credit it does make some reference in passing to the resources. It is a grade-C style answer because it has vague arguments and a set of loosely defined economic factors that ought to have been linked to cities more directly.

■ ■ ■

Question 3

Future food supplies in cities

Explain why there are concerns over future food supplies in cities.   (10 marks)

Time guidance approximately 13 minutes.

Candidate A

There are concerns over future food supplies in cities because of four categories: a rise in prices, an increase in population, changes in diets and an increased awareness of environmental impacts.

A rise in prices causes food to become more expensive and therefore results in a scarcity of food in poorer countries. Food is becoming more expensive because water supply is low (only 0.08% of the world's water supply is available) and water is needed for the production of food, for example it makes semi-arid land useful in production via irrigation. Also the cost of energy is increasing due to limited resources (fossil fuels that supply energy to trucks, boats and planes are dearer because they are scarce). Government taxes and quotas make food more expensive to import.

As population grows — an estimated rise of 3 million by 2030 — the food supply demand becomes a concern. More people will need to be fed and more of the resources will be exploited; we will need triple the amount of food by 2030 according to a BBC study.

Changes in diet cause an increase in demand. As countries such as China become more westernised, their diet changes to a more meat-based diet as opposed to the current staple diet of cereals (wheat, rice, etc.). It takes 7 kg of grain to produce 7 kg of beef, therefore an increased meat-based diet needs more grain, not for the people to eat but for the animals too. Also, the awareness of organic and free-range foods causes changes in food supplies.

The environmental awareness of 'green foods' and opposition to GM crops puts a strain on food supply as well as the emphasis on being self-sufficient in order to reduce the carbon dioxide emissions from importing foods and therefore the carbon footprint.

> The opening paragraph groups several ideas together. However, the four concerns are not linked to city food supplies. The candidate only implies such a link. Always make sure that the links are obvious and, wherever possible, refer to the resources. This is a grade-D/C answer because it is generalised and needs more supporting evidence.

Candidate B

The essential reason why there are concerns over food supplies and indeed water and energy to cities is that urban populations are growing and so is consumption. Figure 7 shows this clearly in that in 2003 $197,818 million was spent on food in China and by 2007 this had risen to a massive $276, 255 million.

As little as 60 years ago just 3% of the world's population lived in cities, but now it is 47%. This means that there are fewer people available to work on the fields in the countryside to produce food. However, some may argue that this is invalid as we are developing new ways to produce food such as vertical farms (Figure 12). This is the Boserup view. We are also using an increased amount of grain to provide fuel rather than food; in the USA a quarter of maize is now used for biofuels.

Figure 16 helps to demonstrate how consumption is greater in more developed countries. As the world's LDCs become more developed and the populations turn to alternative diets, which are often more meat-based and use more resources, food supplies start to become more scarce.

One reason not related to the actual production of food as to why there are concerns over food supply in cities is that it is difficult to transport food into cities. In Shanghai, for example, the average speed on the road is just 16 mph. As the cost of fuel also rises, it is becoming more expensive to get food into city areas as the cost to producers is growing greatly due partly to a rise in fuel costs. Similarly, global warming is causing growing seasons to change and in 2003 the UK lettuces matured too early causing prices to rise later in the season from 30p to 80p due to the shortage.

Sustainable development is development that meets the needs of the present without compromising the ability of future generations to meet their own needs". It contains two key concepts:

- the concept of "needs", in particular the essential needs of the world's poor, to which overriding priority should be given; and

- the idea of limitations imposed by the state of technology and social organization on the environment's ability to meet present and future needs."

The Stool of Sustainability

3-Legged Sustainability Stool

Sustainability

Economic Leg
Good Jobs
Fair wages
Security
Infrastructure
Fair Trade

Environmental Leg
0 Pollution & Waste
Renewable Energy
Conservation
Restoration

Social Leg
Working conditions
Health services
Education services
Community & Culture
Social justice

Quality of Life / Genuine Wealth / Genuine Progress

Defining Sustainable

The Futurity Principle

The development aspirations of future generations must not be impaired by the actions we take today and fot this reason forms one part of the concept known as inter-generational equity or quite simply fair shares for the next generations or descendents. Futurity demands that the value of all assets are passed on to future generations including human knowledge, natural reserves and cultural integrity should not decline.

Some basic guidelines

Renewable resources should not be sonsumed faster than they are produced

Non-renewable resources should not be consumed at at faster rate than they can be replaced by non-renewable resources

Waste resources should not be discharged into the environment faster than it can assiliate them without imparing the environment

The most vulnerable people have access to a satisfactory quality of life particulary with respect to access to resources and development opportunities, with out threat form harm

Some ongoing debates about sustainability

Does a species that has no resource value have any importance?

What is a fair allocation of resources?

> ✍ This answer outlines several concerns and uses the resources. Population increase in cities, consumption in MDC cities, transport issues and climate change and its effect on agriculture are all valid points. This response is on the grade-A/B border.

■ ■ ■

Question 4

Increasing food supplies

Assess the extent to which food supplies can be increased and made more sustainable throughout the world. (25 marks)

Time guidance approximately 33 minutes.

Candidate A

Supplies of food can be increased by adopting alternative forms of supply. One of the alternative forms of supply strategy is vertical farming (Figure 12). Vertical farming is the idea of building a 30-storey building that will provide food for 50,000 people. This supply is sustainable as it has zero net emissions according to the source **www.verticalfarm.com**. However, this might be biased and the construction and transportation of the soil needed would produce emissions so it is not completely environmentally friendly and it is costly. Sustainable development is the balancing act of environmental, social, economic and technological impacts.

Another idea is genetically modified (GM) crops that use science to produce drought-resistant crops, thereby helping LDCs with semi-arid land such as that in Brazil to export goods from previously useless land with the help of GM and irrigation and so help them to rise out of poverty. However, GM causes social conflict as it is not natural and risks have not yet been fully assessed.

City farming works for the Incredible Edible Group of Todmorden, Yorkshire. It aims to be environmentally friendly and self-sufficient in food supply by 2018 as well as bringing the community together (good social effect).

Rooftop farming and 'edible walls' are seen as a more sustainable method as they are on surfaces already made so no energy is used in construction, unlike in vertical farming. They also improve insulation and encourage self-sufficiency of food supply. However, these methods are aesthetic pollution and the types of food that can be grown are limited, for example the UK's climate is not suitable for growing bananas.

These strategies have some negatives such as the cost of vertical farms, the social issues of GM crops and the environmental impacts of importing food stocks, as well as their benefits. I think the only way to make a step towards sustainable development worldwide is to use a combination of all the strategies suitable to that particular country. For example, vertical farms in an LDC are not sustainable as they cost a lot and LDCs would have to rely on government funding, which can go wrong. This happened in the 1960s Green Revolution when only some Asian states funded high-yield variety (HYV) plants and irrigation to improve cultivation substantially; Punjab went from 1.2 tonnes to 4.7 tonnes of wheat yield. Some strategies would be more appropriate in LDCs and some more suited to MDCs due to variations in wealth, development and funding of research. The more sustainable approaches in LDCs would use local materials, local people and increase independence, so migration may be best there.

Different strategies in separate countries that are appropriate and mixed is the answer in my opinion, in conjunction with reducing consumption. For example, every year the average UK household wastes an average of £420 of food; the Love Food Hate Waste campaign is educating people to reduce consumption, which in turn reduces the strain on food supply. Therefore, strategies to increase supply are sustainable steps but they would not be needed so much, and therefore the negative impacts would not be so great or present, if consumption and demand were reduced.

> The candidate addresses a series of alternative strategies for sustainable food supply. The answer notes the fact that different strategies are needed for different places. It looks at both consumption and supply. This essay is reasonably well argued and obtains a grade A.

Candidate B

On the face of things it appears the job of producing more food and making it sustainable is an impossible task due to an estimated 3 billion extra people expected to populate the Earth in the next 50 years (Figure 2). However, if we look at Boserup's view that humankind will discover and innovate new methods of dealing with the world's problems, it appears things are already happening to make our food supply sustainable for the future.

Figure 10 talks about GM crops and how they will be the way forward when it comes to meeting food supply. They could allow for shorter growing seasons, meaning harvest would not have to happen just once a year. Others would adopt the argument that this is unsustainable as it will require increased amounts of seeds for farmers.

In my opinion, and the opinion of many, to sustain food supplies we must locate new areas where we can grow food, preferably reducing food miles at the same time. If we look at the case of southeast England, however, population density is 425 persons per km², a figure that does not provide much room to grow crops in fields. In fact people have been so worried about food supply and the cost of food there is now a 5-year waiting list in some towns for allotments, much like those shown in Figure 13. A possible solution for southeast England would be to grow food in the protected areas around cities, for example around London where land is protected from development under the green belt scheme. This idea, however, seems weak considering the green belt in London is threatened over the human rights of travellers to settle there and the need to build housing in these areas. Similar stories are occurring in the expansion of York University and the expansion of the M6.

The projects and initiatives, however, do not have to be small scale by any means. The second Green Revolution, in which we expect to raise grain production from 214 million tonnes to 400 million tonnes, would provide a vast quantity of food for the growing population. The first Green Revolution was labelled highly successful — in Mexico, for example, they were able to become self-sufficient by 1956 and then started exporting food by 1964. The sustainability of our energy supply, however, greatly affects this due to increased amounts of food being used for biofuels rather than food: in the USA, 25% of maize is used for energy rather than food.

Since sustainable food means producing food that meets the demands of the population without causing harmful negative effects to the environment or future generations, it is worth looking at how we are producing fish. Fish supplies are decreasing rapidly due to intensive fishing. Also our increased demand for water damages ecosystems — for example the River Kennet was used for trout farming but Thames Water sabotaged this as it pumped more and more water from the river. We have therefore taken to a new method of fish farming called aquaculture. This produces one-third of the world's fish supply causing a relatively small harmful effect on the environment, although there are worries it may decrease biodiversity.

Aquaponics and hydroponics are two of the most sustainable methods of modern food production. By using a soil solution rather than soil and controlling every single variable condition, we could increase harvests to multiple times each year, rather than just once. It also allows for the use of grey water, helping to reduce water consumption — another feature of the world's sustainability — and, if used in vertical farms like we see in Figure 12 could reduce the need for trucks and lorries to travel great distances to move food as consumption would happen in the same city, decreasing energy demand and improving energy sustainability.

In conclusion, it appears that there are a variety of ways to improve the supply of food and I am sure we will be successful in this. To improve sustainability, however, we must tackle the uneven distribution of food because, as Figure 15 shows, countries such as the USA and Spain consume 3,200 kcals and above per person while Africans struggle in many cases to consume 2,000 kcals per person. To make food supply sustainable, therefore, we must support LDCs that fail to feed their populations.

> The candidate rightly recognises the opportunity to write a full essay. Although the style is somewhat colloquial, the substance of the topic is covered, again with reference to some of the resources. The essay has a concluding paragraph but lacks a real introduction. Sometimes it digresses, such as in the paragraph discussing travellers. It could have been more tightly argued. Nevertheless, this gains an A grade because of its ideas and factual content.

■ ■ ■

Section B

In this section you may use information from your studies for AS and A2 geography as well as from the Resource Folder and your own research.

Question 5

Sustaining energy needs

'For all countries, future energy needs are not sustainable without a lower standard of living.' How far do you agree? (25 marks)

Time guidance approximately 33 minutes.

Candidate A

Currently, the lifestyles adopted throughout the world are not sustainable. The Bruntland Commission says that sustainable development means meeting the needs of the present without compromising the resources for future generations. For energy consumption to be sustainable, a country's supply needs to outweigh negative economic, political and environmental impacts as well as ensuring the level of technology is appropriate for that individual country. For example, what may appear to be sustainable for one country may not be for another. The main difference of this classification is that of MDCs and LDCs.

MDCs are more equipped and developed, for example the average annual income in Sweden is $25,030, so therefore their standard of living is high. Energy is used in everyday lives and, the more developed that standard of living, the more energy is used. Energy can be consumed via electricity or via transport, etc. The increased standard of living in Western society means more time to socialise with loved ones, and time-saving appliances can be afforded due to increased wages and these require energy. An increased standard of living means people can afford to take holidays; energy is used. The lifestyle is unsustainable as fossil fuels, which contribute to 87% of the energy supply, are finite and many people think maximum oil output has already been reached. This causes a great strain on the environment as new forms of energy supply need to be discovered or demand drastically reduced. For example, the *New York Times* in 2009 released a paper claiming that all the gadgets powered by energy in the USA need the equivalent of 560 coal-fired power plants to power them. Therefore, this leads me to agree that a decrease in the standard of living is needed.

However, Western ways are being adopted in emerging economies such as China and Japan, whose economic growth relies on the growth of energy industry, as well as in extremely poor countries such as Chad in North Africa, which have large oil reserves. Therefore, a reduction in the standard of living, which would decrease energy usage, would negatively affect their economies.

Also, to achieve sustainability, schemes must be increasing the quality of life for all. By denying countries that are emerging the lifestyle that we have causes an unfair moral issue as well as possible political conflict.

I agree somewhat with the statement that all countries need to have a lower standard of living and therefore greater efficiency and a reduction in demand for energy. However, I think this idea needs to be enforced via education not force, like the House Management Project in 1998 in Chad in which the World Bank and non-governmental organisations provided energy-efficient clay stoves (75% reduction). The people were educated on how to manage biomass so it does not deplete and advised on the need for it to be used in conjunction with other energy supplies that are cheaper and more green such as wind power, which makes up 69% of Norway's energy supply. Both a change in supply and definitely a change in lifestyle need to be implemented in my opinion.

 This question was the first one attempted by the candidate in the examination. There is no harm taking that approach if you are more confident about a particular question. The response focuses on the question and argues from a more general and theoretical position. Eventually, examples emerge. The concluding paragraph returns to further new examples, which a conclusion should not do. It is a grade-B performance.

Candidate B

Sustainable energy is producing energy in a way that meets the levels of demand for a population while ensuring the negative side effects are minimal and future generations are not left to clear up the mess that is made. I can understand why people may take the view that we must lower our standard of living to make energy sustainable. This is because in recent years our demand for energy has increased so dramatically through increased income and better lifestyles, allowing for more cars, more heated swimming pools, more electrical appliances and more demand for air travel (40% of the world's goods are now transported by air cargo and 2.3 billion people use air transport, employing millions of people worldwide). All this causes us to consume far more resources as our demand for energy grows. However, as our demand increases, we are finding more ways to produce energy.

There are plans for a £240 billion solar panel project to be built in the Sahara desert. This would provide Europe with 15% of its total energy supply, at the same time as employing thousands of African people to operate the solar panels allowing them to offer their own communities better lifestyles. Similarly, UK offshore wind farms could employ 30,000 UK workers and provide the UK with 30% of its energy needs. However, not all our methods of producing energy need to be on such a large scale. In India, for example, villages can produce 90% of their energy requirements through animal dung. In Ethiopia, small-scale projects mean communities can build their own dams and provide themselves with both energy and water.

Some people are of the opinion that the new methods of energy supply we produce will still not be sufficient. They favour the idea of reducing demand. A UK government campaign of driving 5 miles less a week tackles this exact problem. In Edinburgh there are projects such as the City Car Club. This aims to reduce the number of cars on the roads which, in turn, reduces the demand for fuel and energy. It works by people booking cars for days or even weeks. All the cars are less than 3 years old, meaning their energy efficiency is better and each car in the club replaces 8–10 privately owned vehicles. Although this has decreased energy demand, it has not lowered the standard of living.

Another way that we can allow for future energy demands to be met is to increase energy efficiency. In Athens a car retirement scheme between 1991 and 1992 retired over 100,000 cars and replaced them with newer and more efficient vehicles, at the same time improving environmental quality. If anything, this helped to improve the lifestyles of people who had respiratory diseases caused by energy emissions. Other ways to become efficient would be simple things like not leaving lights on or appliances on standby, none of which in my mind reduce the quality of life.

In conclusion, I do not agree with the statement at all. I believe that energy will be sustainable and, although sacrifices may be made in some cases, their benefits will only improve living standards. For example the savings made on an energy-efficient light bulb may go towards a holiday, increasing living standards. The real question, however, must be, how can we become more energy efficient and sustainable? If we fail to do this, the problems we will encounter will be far greater than those of reducing consumption.

> This is a good-quality answer of A-grade standard. It notes changes in energy demands and introduces a couple of strategies that are aimed at increased sustainability and improving living standards. Once again, this candidate would benefit by spending more time composing the essay rather than rushing everything on to paper. In spite of the deficiencies of style, it does have supporting evidence from a range of countries.

Overall comment

Candidate A

> This candidate has underperformed on the first three responses but recovered with two better responses to the final two questions. In this case, the three weaker responses would have resulted in an overall grade B.

Candidate B

> This candidate has performed well on all but Question 2. However, the overall result would have been an A grade.